THE MAKING OF
GEOGRAPHY

THE MAKING OF
GEOGRAPHY

BY
R. E. DICKINSON
AND
O. J. R. HOWARTH

GREENWOOD PRESS, PUBLISHERS
WESTPORT, CONNECTICUT

Library of Congress Cataloging in Publication Data

Dickinson, Robert Eric, 1905-
   The making of geography.

   Reprint with new pref. and postscript, of the 1933 ed.
published by Clarendon Press, Oxford.
   Bibliography: p.
   Includes index.
   1. Geography--History. I. Howarth, Osbert John Rad-
cliffe, 1877-   joint author. II. Title.
G80.D5 1976   910'.9   75-38379
ISBN 0-8371-8669-2

Originally published in 1933 at the Clarendon Press, Oxford

This reprint has been authorized by the Clarendon Press Oxford

Reprinted in 1976 by Greenwood Press,
a division of Williamhouse-Regency Inc.

Library of Congress Catalog Card Number 75-38379

ISBN 0-8371-8669-2

Printed in the United States of America

CONTENTS

LIST OF ILLUSTRATIONS IN TEXT

LIST OF PLATES

PREFACE

This book has a long history. In 1913, before geography was established as an independent university study, a small book appeared entitled *A History of Geography,* written by (Sir) J. Scott Keltie, secretary of the Royal Geographical Society, and his young assistant, O. J. R. Howarth, a former student of Herbertson at Oxford. In 1929, the Oxford University Press approached Howarth, the surviving author, then secretary of the British Association for the Advancement of Science, to rewrite the book. Howarth invited me to collaborate (on the recommendation of my chief, C. B. Fawcett, also a former Herbertson student). Howarth rewrote the section on the classical and medieval periods, and I wrote two thirds of the new book, beginning at 1500 and ending with the situation in the twenties.

This is a reprint of the same book. Vast advances have taken place over the past fifty years, a history in itself. These may be studied in my two major works, *Makers of Modern Geography* (1967) and *The Regional Concept: The Anglo-American Leaders* (1976), both published by Routledge and Kegan Paul, London. The conceptual framework of this field of study and its practical application are presented in my *Regional Ecology: The Study of Man's Environment* (1971), Wiley, New York) and *Environments of America* (1975, Vantage, New York). The present work is reprinted intact as written in 1930, with the addition of a short postscript.

Robert E. Dickinson.

POSTSCRIPT, 1975
by Robert E. Dickinson

There has been a remarkable growth of geographic scholarship over the past fifty years. This is evidenced by the great increase in the membership of old and new professional organizations; the establishment of independent departments of geography in universities throughout the world; the growth in membership and prestige of the International Geographical Union and the work of its research commissions; the many new periodicals, published in nearly fifty languages; the soaring enrolment of students, both undergraduates and graduates, in the universities; and the teaching of geography, to an advanced level, in schools throughout Europe, with, however, its serious elimination in the school systems of America.

Geography found its peak during the interwar period in Europe in the many volumes, each written by a specialist, of the French *Géographie Universelle,* a regional geography of the world, started by P. Vidal de la Blache, and the similar German *Handbuch der Geogischischen Wissenschaft.* The regional studies of Germany, initiated by F. Ratzel, continue to this day, and now number over one hundred volumes. The massive regional monographs of French geographers (the academic entry to university status), started in the 1900s by Vidal, also continue. Both series change in viewpoint and procedure, but always contend with problems of regionalization. There also appeared in the late forties the three volume work on *Les Fondements de la Géographie Humaine,* an ecological approach, from the pen of M. Sorre, an early student of Vidal. This work does for the geography of man what De Martonne had done for physical geography, landforms, vegetation and climate forty years earlier. The regional study of land and landscape found its florescence of conceptual approach and literary exposition in the works of R. Blanchard in the thirties, eleven volumes on the French Alps, and two volumes on Eastern (French) Canada.

No such mammoth undertakings have appeared in English, but two works mark the culmination of this period. R. Hartshorne's *Nature of Geography* (1939) runs close to the views of Hettner in Germany, but underrates Schlüter and virtually ignores the French contribution. *American Geography: Inventory and Prospect* (1953), edited by P. E. James and C. F. Jones, is a collection of essays by senior scholars on systematic aspects, in which, as one critic put it, they sacrifice their birthright for a mess of potage. Sauer's important work on *The Morphology of Landscape* (1925), though difficult to read because of its Germanic style, remains a cornerstone, but had a negligible impact in Britain and America. Brilliant exceptions occurred among his own students, and among a group of young men in the Middle West in the thirties, the self-styled "chorographers," led by P. E. James, whose culminating work, *Latin America,* is an exemplary exposition of the regional approach in contradistinction to the "environmental determinism" that prevailed until the fifties.

Since 1945 there has been a shift to a more scientific approach, fired by W. Isard's "regional science." Attention has shifted to natural and human processes involved in the location and distribution of spatial phenomena, and in the origin and development of types of earth feature, both natural and manmade. The result is the fertilization of ancillary disciplines, but the neglect, in both research and tuition, of the core of geography, namely, the regional concept.

The regional concept, which receives lip service, but is generally misunderstood, calls for a restatement. It seeks to characterize and explain the uniqueness of terrestrial areas, viewed as associations of earthbound phenomena, physical, biotic, and human, both as 'specific' regions, such as Wales and New England, and worldwide regions, such as landforms, climates, vegetation, economies, and cultures. Such entities are arranged in a mosaic, with cores and wide transitional peripheries, from a local to a worldwide scale. Such study takes priority over the pursuit of the spatial aspects of ancillary disciplines. It uses their relevant skills, while never losing sight of its own skills, approach, and goals.

In the light of this concept, we may now very briefly note the trends in Britain and America since 1945.

In Britain, great advances have been made in every direction; in tuition, research, service, and public recognition. Major contributions have been made in geomorphology, though the label "geochronology" would be a more appropriate title for its labours. Historical geography, pursued with the skills of the historian (especially the medievalist and the archaeologist) would also be better described as "geographical history," with its greatest achievement in the publications of H. C. Darby colleagues on Doomsday (11th century) England. Another field of special importance is on the economic side. W. Smith produced in the late thirties a statistical study of economic regional associations of similar and interconnected phenomena in Britain. (Sir) L. D. Stamp initiated and finally summarized in one big volume a survey of land use in Britain, prepared field by field on the Ordnance Survey map of six inches to one mile, undertaken in schools and universities. This finally appeared as a complete series, published by the Ordnance Survey, on a scale of one inch to one mile. This series was accompanied by many monographs, all prepared by geographers, under Stamp's general direction. The survey is now under revision. Stamp was knighted for his work on "the use of the land."

It is surprising how little attention has been given to the regional variations of landscape. (This does not hold for the continent, where landscape has been embraced since about 1900). In Britain it was virtually ignored between the wars, and indeed such "cryptic study" rejected as unworthy of serious attention. Its pursuit is now in the hands of certain economic historians, with boots and spades, who have enlisted the support of some younger geographers. In America such studies are exceptional. For that very reason, a few outstanding studies should be noted on the American scene—regional types of farmstead, the classification and distribution of hamlets, the spread of folk housing down to 1850, the spread of urban street designs, and the southern plantation system. There is popular concern with environment and ecology, but American geographers have given little attention to the geographic range of urbanization, except for Berry's provocative work on the widening range of metropolitan areas, based upon the (unprinted) sample data of the 1960 enumeration districts.

The study of regional associations as unique and distinctive entities demands not only the assembly, mapping, and accurate geographic delineation of key indicators, but also the evaluation of the changing impact of new techniques, in transport, social provision, and organization on regional segregation and orientatin. Lip service was given in the twenties to the French geographic concepts of *personnalité, pays,* and *paysage,* but nothing was done to investigate the meaning of personality, regional names, countryside, or urban designs. The nearest approaches in the postwar years come from America in the works of D. W. Meinig and A. Clark.

In this connection, some American works must be noted, however briefly. The essays on *Regions of the United States,* edited by J. Fraser Hart (1972), a refreshing switch from the current scientific trend, are truly geographic in the regional sense. Meinig's essay on "The American Wests' offers a stimulating key to historical changes relevant to regional segregation. (With this I would like to couple Borchert's essay on metropolitan development in the United States, published in the *Geographical Review*). Hart seeks to answer the question of the content (indicators) and geographic range (core and periphery) of the Middle West. But the really exemplary regional studies are Jean Gottmann's *Virginia at Mid-Century* (now in a second edition) and *Megalopolis,* sponsored by the Twentieth Century Fund. The first exemplifies the approach of a French geographer, for such is Gottmann's training (Paris). The second work seeks to answer the question of the meaning and significance of the urbanized area of the N. E. Atlantic seaboard, with respect to land use, traffic, organization and attitudes. He uses the skills of various specialists to serve these ends. Here is the geographer's craft put into operation. This has received widespread attention and has even reached a paperback edition in spite of its mammoth size. Gottmann, be it again emphasized, is a French geographer habituated to the American scene.

Geography in Britain is today offered as an honors degree in all universities. The subject has grown in scholastic strength and methodological direction, under the main leadership of P. Haggett. It has received priorities in appointments (with new chairs in both human and physical geography), buildings, equipment, and government aid for research. On the other hand, in the United States, many so-called geographers have become obsessed with the new field of "regional science" and seek to establish their study as a social or behavioral science. A philosophy of universal education prevails, but geography has been virtually eliminated from the school curriculum. Geographers are thus still trying to create an image among students and the wide public, who are still at the "cape and bay" stage which I remember as a small child. In Britain today, the subject is well established as a field of study and as a general education, and it contributes in substantial measure to the teaching profession, business and government.

Chapter 1

GEOGRAPHY IN EARLY CIVILIZATION

THE ancient civilizations which contributed to the making of geography as a department of knowledge belong to the Near East. The geographical instinct, in one form or another, is naturally of early development, and other ancient civilizations must have been possessed of geographical knowledge or ideas. But the achievements in science of the Indian and Far Eastern peoples (for example) who attained high development in prehistoric times are little known, and had no apparent influence upon the West. So that even if the Chinese acquired that very early knowledge of the use of the compass as an aid to land-travel, with which they are commonly but at least doubtfully credited, it is not our task to investigate this and other such questions. The opening historical setting for our study may be very briefly summarized as follows.

In the fourth millennium before Christ there appear the organized states of the lower Tigris-Euphrates and Nile basins: the Sumerians in the one, the Egyptians in the other. Early in the third millennium the Minoan civilization emerges in Crete, and a high standard of culture came to be reached elsewhere in the Aegean area. About 2400 B.C. Assyrian kings were established at Asshur and the first Babylonian dynasty was founded about 2230. Aryan movements to Persia took place from the beginning of the second millennium, and to India from 1600 B.C. In the middle of the same millennium the short-lived Hittite power developed in Asia Minor, as also did the Mycenaean culture in the Peloponnese. Invasions from the lands to the north of the Aegean area supervened about 1200, to overwhelm the civilized states in Crete and the Aegean; at this period also the people of Israel achieved their exodus from Egypt, the

Phoenicians from the eastern Mediterranean seaboard were laying the foundations of their north African colonies, and Babylon passed under the sway of Assyria. In the eighth and seventh centuries before Christ the Assyrian Empire reached the height of its power, and its conquests extended to Egypt. The Phoenicians lost their commercial eminence in the eastern Mediterranean as it was superseded by that of the Greeks, while Phoenician colonies still held their places in the western Mediterranean, and their ships penetrated the ocean northward and southward beyond the Straits of Gibraltar. Greece and Assyria were thus in peaceful contact, a condition which bears upon the history of our subject, as will appear: Greece also withstood the Persians when from 553 to 525 they carried their arms afar over Babylonia, western Asia, and Egypt. This first paragraph of summary political history may fitly end with reference to the conquests of Alexander (338 B.C., and following years), whose victorious armies and fleets, bringing the whole area of Greece, Egypt, the eastern Mediterranean, western Asia, and Persia at least nominally under Macedonian rule, and carrying the force of arms even into India, provided incidentally the most notable series of events in the early story of geographical exploration.

Even in this first period of our study it is not unprofitable to consider the scope of geographical knowledge under three familiar headings: data as to the inhabited world, beliefs as to the figure of the earth and its place in the universe, and measurement. The order of these headings is arbitrary but seems natural and is of no great importance, for they cannot be considered independently without losing the historical thread. Beginning, however, with the first of the three, the chorographical [1] division of our subject: we have ranged in

---

[1] Chorography, describing, description of, districts (more limited than *geography*, less than *topography*) (*Concise Oxford Dictionary*). The word is convenient to our need at this stage. For the use of these terms at a later stage, see Chap. XV.

the preceding paragraph from India to the Atlantic, and we may read of very early contacts between the Aegean and Asia Minor and 'the north' (the continental mass of Eurasia) on the one hand, Africa on the other. Chorographical description must have passed almost wholly by word of mouth; but it must have passed, and perhaps more freely than is realized at first thought. Among traders, however secretive as a class (the Phoenicians, in particular, had that reputation), there must have been talkative individuals; opposing soldiery must sometimes have fraternized.

Let descriptive geography, for our present historical purpose, find its tap-root in poetry, and that of the highest order: no branch of knowledge can more fitly do so. The birth of the school which created the Homeric poems would seem to be assignable to the later part of the second millennium before Christ. The limit in them of any chorographical knowledge approaching exactitude is that of lands immediately bordering the Aegean Sea; but there is some knowledge of lands and peoples to the north; Egypt is known for its wealth, and black men and even pygmies have been heard of in Africa. And as for the story of the wanderings of Odysseus beyond the limits of known lands and seas, is it not, in respect of its descriptions of things seen, of dangers and pleasures encountered, just such as might be based by a poet upon travellers' tales—the tales themselves touched in the shaping, perhaps, by poetic licence, and previously no doubt often embellished in their passage from one teller to another, if not indeed from one language to another? But however vague they may have been, it is not necessary to suppose that they were always distorted beyond recognition. And it seems easier to believe that the poet of *Odyssey* X had heard of the unending summer day on some fiord-coast in the far north when he composed the beautiful description of the Laestrygon's land, than that out of his imagination he hit upon a collection of correlated phenomena which we now recognize so well. So too with the dark distant land of the Cimmerians: may its description not be based on stories of the long winter night in the north? And the Cimmerians themselves, whatever their origin, whether it be Mongol or, as one guess would have it, traceable to Jutland, must for long have had at least indirect contacts with the south, for they became only too well known when in the eighth and following centuries they were invading Asia Minor as far as the Aegean shore.

These examples are put forward to suggest that there existed in the ancient world not, indeed, knowledge, but ideas, however vague, about lands between the Arabian Sea and the Atlantic, the Arctic Circle and the Equator. The persistent and widespread conception of the earth as surrounded by the river of Ocean, very ancient as it is, might even tempt us to speculation as to its possible foundation in scattered facts. But so far as any approach to precision in geographical knowledge is to be supposed, we must conceive of each of the old Near Eastern civilizations as recognizing its own particular little world. Thus a Sumerian map of the world, so-called, is known as having been made to illustrate the military operations of Sargon of Akkad (2700 B.C.), and this obviously represents the Sumerian 'world' as such. It should be added here, however, that the continental division between Europe and Asia must be of pretty high antiquity, if those names are correctly, as commonly, traced to Assyrian words meaning respectively lands toward the setting (west) and rising (east) of the sun. The name of Europe first appears (subject to a questionable reading) in the Homeric *Hymn to Apollo* of the eighth or seventh century B.C., where it distinguishes the mainland to the north from the Peloponnese and the Aegean islands. By the sixth century the division seems to have been well known, though not quite clearly defined (which, for that matter, it is not now).

Early conceptions of the figure of the earth varied only in

detail. Babylonia regarded the earth as the floor of a domed casket, of which the sky was the dome, having its foundations beyond the ocean. The centre of the earth was in the high

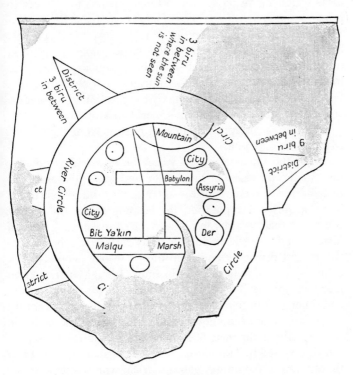

FIG. 1. Sumerian Map of the World, 2,700 B.C.

snow-clad mountains to the north, in which the national river, the Euphrates, has its source. Egypt held a similar view, save that here the world was regarded as oblong, with Egypt as a shallow basin in its centre. It is pertinent to add that in the study of the sky Sumer, Babylon, and Assyria surpassed Egypt. In Egypt scientific development was poorer by contrast with that of art and religion than in the other early civilized states for which there is evidence. The era of Egypt was definitely pre-scientific, and into the ideas which Egypt passed on to Greece it was for Greece to infuse some spirit of science. The states of the Euphrates-Tigris on the other hand, in the same era, inspired that spirit: here, despite all the superstructure of astrology and magic, there is found the origin of scientific astronomy. Here was invented the gnomon, a rod set upright on a horizontal surface, and used to show the time by the position of its shadow and also in observing the sun's meridian altitude.

Crete, so far as appears, did but borrow from Egypt, and Mycenae from Crete.

As for the Phoenicians, they may be regarded as middlemen in the diffusion of knowledge and ideas just as in the distribution of merchandise—so far as they did diffuse them. They acquired a reputation for keeping secret the distant sources of their wealth; and indeed the course of geographical history might have been different if leadership in geographical theory and discovery had been united in one people. They dealt, or are said to have dealt, in amber brought by European traders from the Baltic shores, in tin from Cornwall, in gold and other gifts brought to King Solomon in the tenth century B.C., from sources unknown but usually placed far south in Africa. About the year 600, according to a story preserved but discredited by Herodotus, ships of theirs occupied nearly three years in circumnavigating Africa, which was not done again for two thousand years. Such stories have a way of being founded on fact. Certainly about 500 B.C. one of their fleets reached the Bight of Benin or thereabouts, and another seems to have visited Britain.

A system of land-measurement was an obvious early necessity in level river-plains where landmarks were few, and especially in Egypt where boundaries were apt to be obliterated by the periodical flooding of its valley by the Nile.

Egypt does not appear to have carried geometry farther than this sort of utilitarian practice; but in Babylonia, where a standard system of length and other measures was established about the middle of the third millennium B.C., geometrical conceptions developed ideals, theories, and systems of prophecy connected with numbers, which influenced later thought.

Chapter II

GREEK PHILOSOPHERS AND HISTORIANS

THE Homeric age of Greece, if typified by its extant poetry, was instinct with the spirit of adventure and with an appreciation of scenic beauty by land and sea (characteristics, by the way, both favourable equipment in the making of geographers). By the beginning of the sixth century B.C., the Grecian world had spread from peninsular Greece to the islands and the eastern and northern seaboards of the Aegean, and beyond these, as far as Italy, Sicily, the Rhône mouth and the African shore, on the one hand, the coasts of the Euxine Sea on the other, there were Greek settlements. Some of these, especially the outliers, were purely trading-stations; but others, the more typical, were fully developed city-states. The Greeks were divided into three branches, the Ionian, Dorian, and Aeolian, of which, on the west coast of Asia Minor, the Ionians occupied the central position and were by far the most important. Miletus, well situated at the mouth of the river Maeander, was a place of importance before the Ionian migration, and after the Ionian settlement there it became, during the seventh century, probably the most powerful Greek city, and so continued until the western parts of Asia Minor came under Persian domination toward the close of the sixth century. It was a seat not only of commerce but of culture, and one of its sons is the first individual with whom, among makers of geography, we have to deal.

Thales of Miletus (640–546 B.C.) is regarded as the founder of Greek physical science and philosophy. He was ranked in his own day as first of the seven wise men of Greece, and his fame survived unchallenged for many generations. It was based largely upon his prediction of an eclipse (generally accepted as being that of 585) which brought to an end a battle between the Medes and the Lydians. He was not only a philosopher but a practical man of commerce, and, probably in the second of these capacities, he visited Egypt and there became acquainted with the elementary Egyptian geometry of surfaces already mentioned. Thales, however, introduced abstract geometry to his fellow philosophers and founded the geometry of lines and (omitting all else as inappropriate to our history) he applied his theories to the practice of measuring heights and distances. Speculation did not lead him any nearer than his predecessors to a right conception of the figure of the earth; but as he regarded water as the first element and origin of all things, so he supposed the flat disk of the earth to float in water. Anaximander (c. 611–546 B.C.), a disciple of Thales, conceived as emerging out of chaos a cylindrical mass which was the earth, suspended in a spherical universe. On the practical side, he is said to have introduced the gnomon to the Greeks, presumably from Assyria, and to have made a map of the world on information collected from sailors in Miletus. It is conceived to have been this map or a copy of it, engraved on bronze, which Aristagoras, regent of Miletus, exhibited to the Spartans when asking their help against Persia in 499 B.C.: Herodotus states that it showed 'the circuit of the whole earth, every sea, and all rivers'; and it was so far successful in indicating the distance of Persia from the Mediterranean Sea that the Spartans rejected the solicitation of help.

The Ionian view of the cosmos was not constant: Anaximenes of Miletus, who flourished in the second half of the sixth century, supposed air, not water, to be the source of all things, and the flat earth, in his view, rested on air. Later philosophers of this school were led by Anaxagoras of Clazomenae in Asia Minor (c. 500–428) to an early conception of the atomic theory, and a curious parallel with conditions which have supervened down to modern times is discovered when we learn that his new ideas of the universal order brought him into conflict with established popular religion at Athens where he had settled, and where Greek culture, as the prosperity of Miletus decreased, was finding its principal centre. Nevertheless Democritus of Abdera (perhaps *c.* 470–380, but both dates are very doubtful) more clearly defined ideas as to the origin of the universe through movements of atoms in the void: he supposed the falling of atoms of different weights at different speeds to set up movements which combined to cause rotation of the system. A corollary of this philosophy was the possibility of an infinite number of worlds—so far had the extraordinary speculative activity of the Ionian school led its teachers. But the school from now lost its grip upon Greek thought; and it only remains to observe that like Thales most of its leaders appear to have been travellers—Democritus, for instance, visited both the Euphrates and Nile lands and informed himself as to the geometrical and physical systems as there taught. Travel was almost a necessary preparation for any groundwork of exact knowledge or for philosophical speculation. Democritus constructed a map of the world, revealing a view of it as longer from east to west than from north to south.

Meanwhile Pythagoras, born probably at Samos about 582, is said to have travelled, like other philosophers, in Egypt and the eastern Mediterranean region, and he settled in southern Italy about 529. He founded the Pythagorean school, a moral brotherhood, which so far as concerns its contribution to geography is noteworthy as providing the first known conception of the earth as a globe, revolving around an unseen central fire together with the five known planets, the sun, the moon, the fixed stars, and the antichthon or counter-earth which was necessary to the mathematical theories of the school. The idea of a spherical earth seems to have been due, not to Pythagoras, but to his follower Philolaus (born *c.* 480 in southern Italy). It was based, not upon observation, but rather upon a numerical symbolism: the sphere was con-

sidered to be the perfect figure; the number of the ten revolving bodies was chosen because ten was regarded as the
perfect number, and as only nine were known, the existence
of the antichthon was assumed. The Pythagoreans discovered
the ratios of the length of strings which give a note, its fifth,
and its octave, and they associated a scheme of related
numbers, based upon such ratios, with indivisible units of
space, and held that the distances of the planets from the
earth accorded with a numerical progression: with this
doctrine is connected that well-known phrase 'the music of
the spheres'. At a later period this school of thought evolved
the rotation of the earth about its axis. To the Pythagorean
conception of the universe the author of the Copernican
system, many centuries later, acknowledged his debt; but
the belief in the spherical earth commanded no widespread
adherence at the time, though it continued in a measure to
influence philosophical thought.

We must for a while leave this side of our subject. The
next figure among the makers of geography is again a Miletan
—Hecataeus, who lived about 500 B.C. He was a statesman,
a traveller, and an historian, and has been credited with the
composition of a *Periodos* or circumambulation of the world.
It was supposed to have been in two books, one for Europe,
one for Asia, and to have been accompanied by a map corrected and enlarged from that of Anaximander. The authen-
ticity of this work, however, is very doubtful. It is therefore
impossible to say whether Herodotus owed as much to the
guidance of Hecataeus in his study of geography as he did in
that of history.

Herodotus (*c.* 484–425 B.C.), of Halicarnassus on the west
coast of Asia Minor, came of a family of high standing, and
devoted himself early to literature and to travel.

'He traversed Asia Minor and European Greece probably more
than once; he visited all the important islands of the Archipelago.
. . . He undertook the long and perilous journey from Sardis to

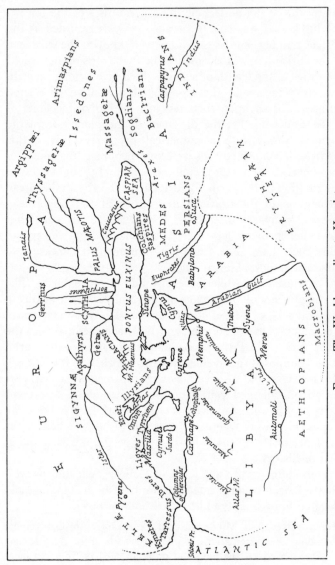

FIG. 2. The World according to Herodotus.

the Persian capital Susa, visited Babylon, Colchis, and the western shores of the Black Sea as far as the estuary of the Dnieper, he travelled in Scythia and in Thrace, visited Zante and Magna Graecia, explored the antiquities of Tyre, coasted along the shores of Palestine, saw Gaza, and made a long stay in Egypt. At all the more interesting sites he took up his abode for a time; he examined, he inquired, he made measurements, he accumulated materials.'[1]

He was an historian primarily, but one with a full sense of the value of geographical setting, in which characteristic he offered a lesson to later historians too seldom accepted by them.

There is no earlier work to compare with that of Herodotus; therefore it is possible only to suggest as a general impression that information based upon hearsay outside the ambit of Greek knowledge and Herodotus' own travels had not materially increased in the previous four or five centuries. In central Europe, for instance, Herodotus knew the Ister (Danube) to be an important river; but he imagined it as rising in a district where we now place the Pyrenees. Of lands and peoples north of Scythia, itself to the north of the Black Sea, he revealed no additional knowledge, and his critical faculty led him to reject fables, though we may now, as has been suggested, suspect some of these of being based upon fact. He had heard something of the tribes of west-central Asia, and of the river Indus. He adopted a third continental division already extant, which distinguished Libya (Africa) from Asia, but he struck a modern note in preferring the Arabian Gulf (Red Sea) as the boundary between them, rather than the Nile as others held, because he preferred not to allow a river inhabited on both banks by the same people to be regarded as a boundary. He imagined the Nile to rise in western Africa and to flow east and then

[1] Rawlinson and E. M. Walker, s.v. *Herodotus*, in *Encyclopaedia Britannica* (11th ed.).

north, in a certain symmetry with the Danube in Europe. He had heard of a great river in western Egypt—whether the Bahr el-Ghazal, or even the Niger, it is useless to conjecture—through the story of five young men who travelled many days southward across the desert from the shore of Syrtis, were captured by small black men by whom they were conveyed to this river, and were afterwards released. Here is just such a story as, being founded on fact, would readily pass from mouth to mouth for the wonder of it. The diagram conveys an idea of Herodotus' geographical range, which justified his belief, according to his lights, in an earth longer from west to east than from north to south, as also Democritus of Abdera had it.

Two-thirds of Herodotus' great work is introductory to his main purpose, which was to record the Persian invasion and the Greek triumph against it. The Persian attempts against Greece in 490 B.C., under Datis and Artaphernes in the reign of Darius I, and in 480–479 under King Xerxes himself, were abortive, and inspired Greece to a sense of nationality and of moral well-being which raised Greek culture to its summit. Geographical conditions, however, were against political unity. The Peloponnesian wars in 460–454 and 431–404 were followed by the brief period of the Spartan supremacy; the Corinthian war of 394–387 by the failure of Thebes' attempt to maintain the leadership of the Greek states. Plato, the Athenian philosopher (427–347 B.C.), experienced the political dissociations of Greece, and could be led by consideration of them into that curious excursion into what is to-day called human geography, when he blamed the sea for its influence upon men, making them unfriends and faithless toward their fellow citizens and neighbouring states. But the rivalry of the city-states of Greece helped the Greek character in its readiness to pursue fresh lines of thought. Plato's philosophy, ethical, intellectual, and mystical as it was, did not carry him far along the

lines we are following here; but his mysticism led him to
record and make great use of one of the most remarkable
known geographical fables—that concerning the lost continent or island of Atlantis. This conception of an unknown
western land first appears in Plato's *Timaeus*. He ascribes it
to an Egyptian source: it has been suggested that it may have
represented Egyptian tradition about Crete in the Minoan
period, and was related to the Homeric description of the
Phaeacians' island in *Odyssey* VIII. By Plato's time the
existing knowledge of the western Mediterranean and the
Straits of Gibraltar had pushed Atlantis, so to say, into the
ocean beyond, and the Atlantic Ocean remained a repository
of this or other similar mythical lands, according to stories
current in various literatures from Arab to Welsh, down to
the eighteenth century of our era. Atlantis, as a geographical
fable, may rank alongside the incident of the Flood—save
that in the case of Atlantis whatever facts the story may have
been based upon are lost, whereas, for the occurrence of the
Mesopotamian flood at least, some evidence has been found.

During the lifetime of Plato there took place (401–400) the
retreat of the Ten Thousand under Xenophon (*c.* 430–350).
These were a band of Greek mercenaries in the service of
Cyrus the Younger, the turbulent son of Darius II of Persia.
Cyrus, from Sardis in Lydia, launched an attack upon Babylonia and was there defeated, whereupon Xenophon and
his men made their way northward through the Armenian
mountains to the coast of the Black Sea at Trapezus (Trebizond) and thence along that coast to the Bosphorus. Xenophon, who in addition to a taste for adventure was a cultivator
of philosophy (in the Socratic school) and history, acted in
the spirit of a military leader with literary taste and made the
best of a bad business by composing his *Anabasis* (which
has been translated as 'An Up-country March'). In this he
described with accuracy things seen in the course of the
journey, and must have given precision to Greek knowledge

of the geography of the country traversed. Perhaps within his time there was written an historical-geographical survey of the world by Ephorus of Cyme in Aeolis (*c.* 400–330) of whose work, however, little is known. There also falls within this period (*c.* 350) a Periplus ('round voyage') or description of the coasts of the Mediterranean Sea, which, whether it was an edition of an earlier work or not, is certainly not, as it stands, that of the author to whom it came to be attributed, Scylax of Caryanda in Caria, who had lived about 500 B.C., and made a land journey to the Indus and returned thence by sea. The India of popular belief seems at this time and after to have been a wonderland, and Ctesias of Cnidus in Caria (fifth century B.C.), a physician at the Persian court and an historian, wrote a description of India which from the surviving summary appears to have given full rein to his imagination. Studies attributed to him on rivers and on mountains are lost; but the indication that geographical work had now reached the stage of general treatises on individual phenomena should be recorded.

Aristotle (384–322 B.C.) was the son of a physician to the Macedonian court, and in his eighteenth year became a pupil of Plato at Athens. From the age of thirty-eight to fifty, however, he travelled and lived elsewhere in the Aegean area, and his work was not confined to philosophical contemplation, but afforded him practical experience both scientific and political, effectively equipping him as a great leader of the philosophical school which in 334 he set up in the Lyceum on his return to Athens. It may be true that of all the departments of knowledge in which he laboured with such wonderful energy and insight, Aristotle himself was least successful in that of geography. But any failure in interpreting geographical facts is offset by his work in accumulating them, and in placing this science, like others, on a basis of systematic observation. He left no specifically geographical work: that on the universe (*Peri Kosmou*) later ascribed to him is now recognized as not his; but in the treatise on the heaven (*Peri Ouranou*) and that called *Meteorologika* he is much concerned with geographical phenomena on the cosmical and physical sides. He accepted the doctrine by now common in Greek philosophy though not within popular understanding, that the earth is a sphere, and he adduced proof of this by laying down that matter attracted by gravity toward a centre would necessarily coalesce in a spherical form; also by showing that the shadow of the earth upon the moon during an eclipse (the mechanism of which he understood) indicates always a circular shape. He distinguished between the inhabited world as known to him, and the inhabitable world. To the first, like predecessors of his already mentioned, he assigned a length greater from west to east than from north to south, indicating incidentally that the popular conception of the inhabited world still made it round. Aristotle placed the earth in the centre of the universe, and supposed it to be unmoving, while the other heavenly bodies revolved around it. He argued its small size from observations upon fixed stars as viewed from different points. He accepted a measurement, however, already current, of 400,000 stadia for its circumference. This would equal about 46,000 miles: the actual round figure at the equator is 25,000 miles. He realized the division of the earth into zones according to temperature, and postulated a southern temperate zone corresponding to the northern temperate zone in which lay his own inhabited world—for he supposed the cold and hot zones respectively north and south of that to be uninhabitable by reason of their temperatures. His conceptions of meteorological phenomena led him so far as to realize that meteorological relations between the temperate and arctic zones in the northern hemisphere would be paralleled by those between the temperate and antarctic zones in the southern, as exemplified by comparison between the cold northerly winds in the one and cold southerly winds in the other. He

made close observations in relation to meteorology and seis-
mology within his own range, and was cognizant of such
changes in land-forms as are effected by the deposition of
alluvium by rivers. His faith in contemporary knowledge of
the outer circle of his inhabited world betrayed him when he
sought for examples in illustration of some of his doctrines,
as, for instance, that which correlated the sources of great
rivers with high mountains. Here reappears the Herodotean
belief in the river Danube rising in the Pyrenees; while
Aristotle's data concerning great rivers rising in the Caucasus
and mountains of central Asia are almost wholly erroneous.
There is evidence that the *Meteorologika* was composed
before the expeditions of Alexander the Great lightened the
darkness of ignorance about Iran, north-western India, and
western Turkistan.

It is interesting, however, to discover in Aristotle himself
one of the sources of inspiration of Macedonian imperialism.
In 343 Aristotle had acted as tutor to the boy Alexander
(356–323 B.C.), and is said to have composed treatises for his
instruction on the duties of a monarch and colonization.

Chapter III

ALEXANDER AND THE FOUNDATION OF ALEXANDRIA

A GREAT part of Alexander's campaigns was of the nature of geographical exploration under arms. Even his early expedition (335) when he led an army not only across the Balkan Mountains but also across the lower Danube, scattering the northern tribes which would have opposed him, was not without geographical importance both in itself and because it was followed by embassies from distant peoples, as far as western Europe, with whom the Greco-Macedonian world had previously had little or no direct contact. Alexander's conquests in Greece itself do not geographically concern us; but the campaigns which occupied him from 334 until his death added largely to topographical knowledge, and long provided authority for future students of geography. His devious route through Asia Minor, even, must have revealed new topographical facts. His incursion into Syria and Egypt is of little importance to us; but it falls to mention that he founded the city of Alexandria, the subsequent fame of which as a centre of geographical as well as other learning was in the true Alexandrian tradition.

In Babylonia and Media (Persia) Alexander was covering ground of which Greek knowledge previously was far from exact: and when he passed eastward from the Caspian Sea into Bactria, when he penetrated the fastnesses of the Hindu Kush, saw cities which afterwards became almost mythically famous such as Balkh and Samarkand, and reached the banks of the Jaxartes (Syr Daria); when, entering India, he ranged along the Indus valley to its mouth; when he pushed westward through the desolation of Iran toward Babylonia again, and when his fleets navigated the Arabian Sea and Persian Gulf—then indeed he was breaking new ground and substi-

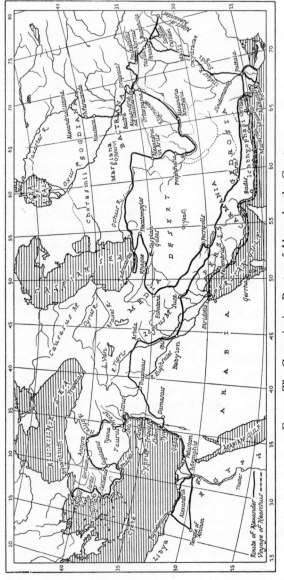

FIG. 3. The Campaigning Routes of Alexander the Great.

uting a much nearer approach to exact knowledge than the vague data which had been at the disposal of Aristotle, Herodotus, and the rest. Nor was it a mere accident of military requirement that he did this. His staff included a civil side, with such observers as Aristobulus of Cassandreia, the historian, who wrote an account of the countries traversed, mainly in the geographical and ethnological aspects; and his bematists or measurers by pacing afforded material for an approach to accuracy in the distance of his marches if not in their direction.

This last weakness emerges in the work of Dichaearchus of Messana (died early in the 3rd cent. B.C.), a pupil of Aristotle, who for the first time, so far as is known, drew a map on a parallel along the Mediterranean Sea, and continued the line along the Taurus and Himalayan ranges. If in alining these ranges he erred, that is no matter for wonder or blame. It was probably he who introduced a lower and therefore nearer estimate of the circumference of the earth, making it about 33,500 miles; he wrote a description of the known world, studied physical geography, and estimated the heights of mountains. He also made a topographical study of Greece, with maps. His works have survived, if at all, only in fragmentary form. Theophrastus of Lesbos (c. 372–287 B.C.), another pupil and closer intimate of Aristotle, carried on the master's work in meteorology, studied rocks and soils, and is most noteworthy for his work in botany. The new material acquired during Alexander's campaigns was accessible to him, and he made use of it by methods which have merited the admiration of botanists ever since; for our purpose it is sufficient to note that he did not fail to appreciate the importance of the study of distribution: we here record for the first time the emergence of plant geography as a department of our subject. Theophrastus by will left his house to the school, with directions as to hanging maps in the colonnade.

At or about the time when Alexander was extending geographical knowledge eastward, Pytheas of Massilia (Marseilles) was doing the same to the west; but as a lone traveller not a military conqueror. He was a navigator with a training in astronomy, who calculated the latitude of Massilia very nearly, and was well equipped, according to his lights, as a scientific explorer. He visited Britain, and heard of land to the north of it, of which he recorded the name of Thule; he also heard of, if he did not visit, the Baltic Sea. He was aware of the prolongation of daylight in summer and darkness in winter in the far north, and his astronomical knowledge enabled him to account for these phenomena. Like so many other voyagers who have asked for information about places beyond their own reach, he was not an implacable critic of the marvellous—nor, if it comes to that, were some of his contemporary travellers in the East. Subsequent students of their results, therefore, were prone to scepticism about some of their statements, right or wrong, and as Pytheas' work is known to us only at second hand we have probably less than a fair view of his attainments. In summary of the advancement of geography as a science in this, the fourth century B.C., it may be said that it was great, and the incompleteness of original records may justify the belief that such workers in the geographical field as Dichaearchus, Theophrastus, and Pytheas were worthy of a higher rank than came to be accorded to them.

The empire of Alexander was divided between his generals after his death; but he had laid down main lines of communication, and thus it was, for example, that Megasthenes the Greek envoy of Seleucus of Babylon came in 303 to the court of the ruler of northern India, Chandragupta Maurya, and by his residence at the capital city on the banks of the Ganges made himself a (not impeccable) authority for Indian descriptive geography upon whom later Greek geographers long relied.

Ptolemy I, later called Soter, the saviour, founded in

Egypt the Macedonian dynasty which continued for three centuries and a half. A more or less peaceful penetration of the country ensued, by Greek officials and colonists, and Alexandria, the conqueror's own city, became a centre of Hellenism and of learning unrivalled until its decline set in with that of the Roman Empire in the third century A.D. The foundation of its high status was laid by Ptolemy Soter himself, who was a patron of letters and founded the great library for which the city became world-famous. With this period of our subject there also synchronizes the rise of Rome, which was the dominant city in Italy by the first half of the third century B.C., made alliance with Egypt in 273, and became a subject of interest to Greek students and visitors.

At this time the ambitions of Rome were in the western rather than toward the eastern Mediterranean area, and so continued until the power of its great rival, Carthage, was broken in 202. But such considerations as the threat from Macedonia in collaboration with Carthage forced Rome to look eastward also, and by the early part of the second century a species of Roman protectorate was spreading throughout the eastern area. Macedonia became a province in 148, and Greece came under direct Roman rule in 146. Egypt and the eastern provinces were not conquered until 30–29 B.C., and throughout the period the older Hellenistic culture maintained itself, but, so far as concerns our subject (and others), with a difference. There appears the natural tendency to criticize earlier beliefs; to require their proof by observed facts when the means of ascertaining those facts were wanting. For example, when Aristarchus of Samos (c. 250 B.C.), after at first accepting the standard doctrine that the earth is the centre of the universe, later held that the earth revolves around the sun, there were scarcely any to accept this anticipation of the Copernican theory, and it was neglected by later investigators. Aristarchus is said to have improved upon the gnomon by setting a rod upright in a bowl, the length of the rod and the radius of the bowl being equal. Angles of altitude were read on circles inscribed inside the bowl. This was the instrument known as the scaph.

The centre of our study, then, is shifted now to Alexandria. The immense advancement of mathematics by such masters as Euclid and Archimedes had as a side-issue an advance in mathematical geography, due to Eratosthenes (c. 276–194 B.C.), chief librarian at the Egyptian capital. He was a man of many interests and wide erudition, whose nickname Beta, however, is suggestive in more than one implication. But if geography be worthy of an honours school, he was worthy of the mark of alpha in it. He wrote a critical history of geography from the Homeric period. He calculated the inhabited world to measure about 9,000 miles west-and-east and 4,400 miles north-and-south; of which the first figure is a third too long for the world he actually knew, but the second is not far from the truth. He mapped his world, drawing an equatorial parallel through the Straits of Gibraltar, six other parallels, and seven meridians at unequal distances apart. He learned that at Syene (Assuan) a deep well was fully illuminated by the sun at the summer solstice; he assumed the place therefore to be on the tropic. He took Alexandria to be on the same meridian, and measured the zenith distance of the sun there also at the solstice. He reckoned the distance between the two places to be 5,000 stades (about 500 geographical miles) apart, and from his observation he took this distance as one-fiftieth of a great circle. Thus he gave the earth a circumference, in round figures, of 25,000 geographical miles, instead of 25,000 statute miles: [1] the truth, as compared with earlier estimates, is seen to be much more nearly approached. His data were inexact. At least, however, in his geographical work he seems to have used his material wisely, and to have recognized its limitations.

[1] This expresses the position conveniently; though there are in fact different views of the length of the stade.

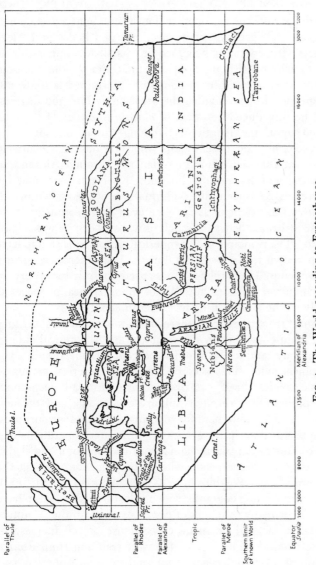

FIG. 4. The World according to Eratosthenes.

Hipparchus the astronomer flourished about 140 B.C., and worked at Nicaea his birthplace, and in Rhodes and Alexandria. In geography he aimed at what it was impossible as yet to achieve with any approach to accuracy—the mapping of the known world upon points of determined latitude and longitude. But he divided the great circle into 360 degrees, and he introduced a system of *climata* or zones of latitude, based upon the length of the longest day on successive parallels, and on observations upon various constellations. The fact that his prime meridian ran through Alexandria, Rhodes, and Byzantium illustrates his limitations, and his inferences concerning the outer margins of the known world, such as the extent of Asia, were even more erroneous than those of Eratosthenes, of much of whose work he was a hard critic. The principle of the spherical astrolabe is said to have been known to Eratosthenes, but the instrument was greatly improved by Hipparchus. It was used not only for observing astronomical altitudes but also for topographical purposes as lately as the sixteenth or seventeenth century A.D.

Apart from these men, there appears in the second century B.C. a shift away from the cosmical and theoretical side of geography, and a bias toward the chorographical and topographical side. This accorded with the practical genius of Rome, which itself contributed very little to geography, except indirectly through military operations, and through survey and measurement for road-making and other purposes. The use of descriptive geography in military operations and in administration was recognized, as was that of a knowledge of distances in connexion with communication through the empire. Thus Polybius, the Greek historian (*c.* 204–122), wrote a book on geography which is lost but is known to have contained many estimates of distances through the Mediterranean area; and he made especial use of the increase of topographical knowledge in the west resulting from Roman conquest. Artemidorus (*c.* 100), said to be an Ephesian, and an extensive traveller, was also a measurer. Agartharchides of Cnidus (*c.* 150), who wrote of the Red Sea, Asia, and Europe, was interested in the geographical environments of peoples. There was also at this period, perhaps under the imperial inspiration of Rome, speculation as to the extent of unknown lands, and Eudoxus of Cyzicus (*c.* 130), navigator, had the true spirit of the explorer, for he not only believed in the possibility of circumnavigating Africa, but attempted it. But of real knowledge of Asia and Africa there was little extension at this period: commerce was established on defined lines, and the military arm of Rome did not reach so far.

Posidonius (*c.* 130–50 B.C.), the Stoic philosopher, travelled and carried out scientific observations in the western Mediterranean area; his philosophy led him to an interest in such phenomena as earthquakes and volcanoes, and he had a clear conception of the influence of the moon, and its position in relation to the sun, upon tides. He made an estimate of the circumference of the earth less near the truth than that of Eratosthenes, and below it, and as his philosophical views held wide currency among leading men in Rome, such errors as this persisted. His estimate was less than three-quarters of that of Eratosthenes—about 18,000 geographical miles.

Marcus Vipsianus Agrippa (63–12 B.C.), the Roman general and statesman, was also a writer on geography among other subjects, and supervised the survey of the empire which had been conceived by Julius Caesar. The results of this survey were shown in a map engraved in marble and set up in a public colonnade; and we hear also of other maps displayed for the public instruction.

Chapter IV

STRABO AND EARLY LATIN WRITERS

STRABO (c. 64 B.C.–A.D. 20), Greek historian and geographer, of Amasia in Pontus, is a great figure in the history of geography, but in some measure by accident. It is necessary to bear in mind that the work of the majority of writers hitherto discussed remains to us only in fragmentary forms or in quotation by later authors. Strabo's *Geography* survives almost intact: only the seventh out of seventeen books is incomplete. It is, in truth, 'an encyclopaedia of information concerning the various countries of the inhabited world as known at the beginning of the Christian era; it is an historical geography; and . . . it is a philosophy of geography' (introduction to H. N. Jones's edition of the text, and translation, 1917). But it is a work which must be judged upon its defects as well as its merits. There is perhaps a temptation towards distaste for the complacence with which the writer regards his own achievements as a traveller and as a student of geography and criticizes the work of his predecessors upon topics on which, as we now know, they showed better judgement than he. Herodotus, Pytheas, Eratosthenes, and others at one point or another come undeservedly under his censure. His own choice of authorities was neither impeccable nor exhaustive, and on the side of mathematical geography he was definitely ill equipped.

Strabo came of a family of means, and was able to devote his life to literature and investigation. He studied at Rome among other places, and at Rome one of his teachers, Tyrannio the grammarian, was also a geographer. Strabo visited Rome several times, and his travels covered parts of Italy, Greece, and Asia Minor, and Egypt as far south as Assuan, while he spent more than five years in Alexandria, no doubt accumulating much of his material there. This would help to account for his very practical use of Roman sources of information. Nevertheless, the distribution of material among his seventeen books affords an idea of the knowledge of the inhabited world available in his time. After two introductory books, Spain and Gaul are dealt with in two, Italy in two, northern and eastern Europe in one, Greece in three, Asia in general and the 'Far East' in one, Asia Minor in three, Persia and India in one, the Euphrates-Tigris region, Syria, and Arabia in one, Africa in one. His main interest lay in political geography and all that that implies. Therefore he described the physical features of countries and their inhabitants in such manner as must have been very useful for reference. But there are limitations worthy of the nineteenth century A.D. in his view of the scope of geography as illustrated by examples taken almost at random from his second introductory book:

'Now as for the matters which he regards as fundamental principles of his science, the geographer must rely on the geometricians who have measured the earth as a whole; and in their turn the geometricians must rely upon the astronomers; and again the astronomers upon the physicists. (Book II. 5. ii.)

'If the country that lies under the equator is temperate, as Eratosthenes says it is (an opinion with which Polybius agrees, though he adds . . . that it is the highest part of the earth, and for that reason is subject to rains, because at the season of the Etesian winds the clouds from the north strike in greatest mass against the mountain peaks in that region), then it would be better to regard this as a third temperate zone, though narrow, than to introduce the two zones beneath the tropics. . . . But Poseidonius objects . . . that there can be no high point on a spherical surface because the surface of a sphere is uniform all round . . .' (after quoting contradictory passages from Poseidonius as to the existence of plains or mountains under the equator, Strabo continues): 'Now here the lack of consistency is obvious; but even if it be admitted that the country beneath the equator is mountainous, another inconsistency, as it seems, would arise; for these same men assert that the ocean is one continuous stream around the earth. How, pray, can they place mountains in the centre of the ocean—unless they refer to certain islands? But however this may be, it falls outside the province of geography. . . .' (II. 3. iii.)

'Now Pytheas of Massilia tells us that the parts about Thule, the most northerly of the Britannic islands, are the farthest north, and that there the circle of the summer tropic is the same as the arctic circle. But from other writers I learn nothing on the subject—neither that there exists any island by the name of Thule, nor whether the northern regions are habitable up to the point where the summer tropic becomes the arctic circle. I consider that the northern limit of the inhabited world is much farther south than where the summer tropic becomes the arctic circle. For modern scientific writers are not able to speak of any country north of Ierne (Ireland), which lies to the north of Britain and near thereto, and is the home of men who are complete savages and lead a miserable existence because of the cold; and therefore I consider the northern limit of our inhabited world is to be placed there.' (II. 5. viii.)

'In the regions about 6300 stadia distant from Byzantium and north of Lake Maeotis (Sea of Azov), the sun attains in the winter days an elevation of six cubits at most, and there the longest day has seventeen equinoctial hours. Since the regions beyond already lie near territory rendered uninhabitable by the cold, they are without value to the geographers.' (II. 5. xlii–xliii.)[1]

To whatever use Strabo's *Geography* may have been put as a work of reference by those to whom it was particularly addressed, it seems to have received no immediate attention from other writers. Neither Pliny nor Ptolemy makes reference to it, and it does not appear to have gained a wide reputation until long after their time—from the fifth century onward. Strabo wrote it late in life; so much is clear. Whether he wrote it in Rome or in Amasia has been argued; but the suggestion has been made that from Amasia it would have been the more likely to fail to make its appeal.

Rome during the ensuing period was consolidating its

[1] Trans. H. L. Jones, with slight variations.

empire in Britain and in north-west Africa, but there was no great expansion elsewhere of geographical knowledge through political events. The small work of Pomponius Mela *De Situ Orbis*, composed probably about A. D. 42, is of interest as the first known purely geographical treatise in Latin and the only òne of the classical period excepting the geographical books of Pliny's *Historia Naturalis* (referred to below). Mela was a native of southern Spain, and his details improved upon those of the Greek geographers only in respect of the west, as in respect of the west coast of Europe, and a vague idea of the position of Scandinavia, which, however, he assumed to be an island. He furnishes, however, the first known definite reference to the Orkney Islands, and, turning to his views on the general geography of the world, which otherwise added nothing to previous ideas, we find there (and nowhere else) an assertion of the habitation of the southern temperate zone by the Antichthones, people separated from the known inhabited world by the torrid zone, impassable by reason of its heat—a curious throw-back to, and amplification of, an idea which does not seem to have caught the fancy of geographers since Aristotle.

Seneca the younger (*c.* 3 B.C.–A.D. 65), among his wide range of writings, composed in his *Naturales Quaestiones* a popular study of physical investigations, including astronomy and meteorology as well as the major features of physical geography. The subject seems to have had some attraction for the educated Roman of this period. Pliny the elder (*c.* 23–79) devoted three books and part of another of his *Historia Naturalis* to geography, but it is poor stuff, and adds nothing of moment to our short survey, which next takes us back to the east. About this time there appears an example of what may well have been a relatively common form of text-book. This is the Periplus of the Erythraean Sea, covering the Red Sea and the Arabian Gulf, and referring briefly to distant points in India, Ceylon, the mouth of the Ganges, and even China (in an indication of land-routes thereto). So far as concerns the nearer parts of the area covered, details are given fully and with knowledge, with especial reference to trade, and it would seem that the work was designed for the guidance of traders in somewhat the same manner as the Admiralty Pilots are designed for the guidance of sailors. This is a fragment of the geographer's raw material on the descriptive side; both Pliny and Ptolemy (who follows) used such works, though Ptolemy apparently had not this particular example.

Plate I.   Portion of the Peutinger Table (from F. C. de Scheybe, *Peutingeriana Tabula Itineraria . . .*, 1753)

Chapter V

PTOLEMY

THE cultivation of geography developed and receded upon a curve (so to say) so well defined, and the original sources available for judgement are so incomplete, that it has seemed good not to claim parenthood of the subject or of any branch of it for any particular writer. We have refrained even from calling Eratosthenes, as he has been called, the father of scientific geography, and we certainly refrain from regarding Marinus of Tyre, as also he has been regarded, as the founder of mathematical geography. Nothing is known of his work save that which is preserved by Ptolemy, who avowedly built upon it. Marinus lived in the second century A.D., made use of earlier students' and travellers' results, and endeavoured to improve maps by care in the estimating of latitudes, longitudes, and distances—without much judgement or success as soon as he got away from well-known regions. Marinus, and Ptolemy after him, rejected Eratosthenes' estimate of the earth's circumference, and adopted the less approximate estimate of Poseidonius.

Claudius Ptolemaeus, commonly known as Ptolemy, mathematician, astronomer, and geographer, was born in Egypt and worked at Alexandria. The dates of his birth and death are not known, but he made his series of astronomical observations from A.D. 127 to 151, and his geography (*Geographike Syntaxis*) is assigned to the period 150–60.

Not in geography only, he was influenced by the work of Hipparchus, and in geography he tried to build upon Hipparchus' fundamental conception that a map should be based upon points of which the latitude and longitude are known. Marinus, the immediate predecessor of Ptolemy, had worked on the same lines: Ptolemy took his results (with due acknowledgement), co-ordinated, and to some extent corrected them.

His work is fundamentally a course of instruction in the making of maps, and a collection of material for doing so. But as with Hipparchus, so with Ptolemy: 'his theoretical science outstripped his power of applying it practically.' After the two introductory books of his geography, the remaining six contain the latitudes and longitudes of some eight thousand places. It is necessary to realize the implications of this. Let us suppose ourselves devoid of instruments of precision, whether for measuring angles or for telling the time. We have an approximation to one or two meridians and parallels. We know pretty well from frequent experience and averages of time occupied in travel the distances between certain points. Beyond these, our data become progressively more vague, until we rely upon a single statement that point B is so many days' sail or march beyond point A, in a more or less indefinite direction. We attempt to apply these data to the fixing of positions—what could be expected of our results? Certainly no more than Ptolemy achieved: but the fundamental difference is that science has no need of such methods now; Ptolemy and his predecessors, longing for scientific precision, confused the appearance of it with probabilities or something less; we may praise, commiserate, or blame them as we will.

Ptolemy held to the theory of the fixed earth, admitting neither its revolution nor its rotation. He accepted, as we have seen, the measurement of its circumference as laid down by Poseidonius—18,000 geographical miles. He adopted Hipparchus' division of the equator into 360 parts (or degrees as they came to be called). His degree, therefore, of longitude at the equator and of latitude was one of 50 miles instead of 60. This implies, for instance, that along any meridian he would have mapped two points 10° apart if he learnt from his authorities that the distance between them was 500 miles; but actually he would be showing the distance as 600 miles. The equator, again, was placed too far north, since its position was reckoned from the known northern tropic taken to pass through Assuan. Further, Ptolemy followed Marinus in choosing an arbitrary prime meridian passing through the Fortunate Isles, which represent the ancient indefinite knowledge of Madeira and the Canaries,

FIG. 5. Outline of Ptolemy's World.

and these they supposed to be about 7° more easterly than the real position.

But one still standard geographical conception had now been reached: that of the network of meridians and parallels. The use of these terms appears first, so far as known to us, in Ptolemy's work; moreover, 'the methods by which he obviated the difficulty of transferring the delineation of different countries from the spherical surface of the globe to the plane surface of an ordinary map differed little from those in use at the present day'. It is clear that Ptolemy's work itself was accompanied by maps. How much those which accompany existing manuscripts have been altered or received additions cannot be estimated. For the original work came into fairly wide use between Ptolemy's own time and the Dark Age; moreover, as will appear, it became one of the foundations of geography at the medieval Renaissance.

As an example of the errors resulting from Ptolemy's attempt at exactitude in positions, there may be instanced the parallel of 36° N. in the Mediterranean area, which was taken to pass through not only the Straits of Gibraltar and Rhodes, as approximately it does, but also through Sardinia and Sicily, while Carthage and that part of the African coast were placed south of it, not north as they are. The line was carried eastward along the supposed course of the Taurus and Himalayan ranges. Another large error even in the well-known Mediterranean area resulted from Hipparchus' estimate that the latitude of Byzantium and Massilia was the same. Again, in dealing with western Mediterranean lands Ptolemy showed a want of knowledge which might have been obtained from Roman road-itineraries: the absence of liaison between Rome and Alexandria is still apparent. Some errors of long standing were corrected; notably the opening of the Caspian Sea northward into the ocean, which according to Herodotus and Aristotle did not exist, but strangely reappeared according to their successors. There is evidence that a good deal of information as to the unknown lands toward the Far East and toward central Africa had become available in the second century A.D., but it seemed to have reached Marinus and Ptolemy in much the same way as we imagined stories of distant lands passing from one teller to another in the early period of our survey. The silk trade with China may be taken to have had such a result. Any knowledge approaching exactitude (according to the standard of the period) concerning central Asia still had not extended beyond the river Jaxartes (Syr Darya) to which Alexander carried it; but Ptolemy had some idea of mountains branching from the Pamir northward of the Himalaya (Tian Shan). Again, since the compilation of the Periplus of the Erythrean Sea, western sailors had further navigated the Bay of Bengal, and brought records for Marinus to misinterpret. We need not go farther than to indicate the vast island of Ceylon (Taprobane), of which this conception of the size had persisted since the time of Alexander, and may have led to the curious neglect of peninsular India. There is believed also to have been confusion between Ceylon and Sumatra. The Golden Chersonese cannot be identified otherwise than with the Malay Peninsula; the gulf beyond it with the Gulf of Siam. In Africa Ptolemy introduces us to the famous Mountains of the Moon, and has received hearsay of lakes in the region of the sources of the Nile. In west Africa there is word of rivers and coastal features but of nothing more than their bare existence; so that there has been occasion for ample disputation as to whether he had heard of the Niger, the Senegal, and Cape Verd, which on his own plotting might be suspected, or whether by errors in plotting he allocated impossible positions to known features north of the Sahara. The British Isles, if not their orientation, are reasonably known, but nothing of the Scandinavian peninsula.

The early belief in the ocean, surrounding the world, doubted by Herodotus and rejected by Hipparchus, was rejected also by Ptolemy. Not only so, but following Marinus he went to an opposite extreme. The stories of further Asia and Africa may well have suggested wide lands unknown. At any rate Ptolemy conceived Asia, in full continental bulk, extending beyond his own meridian of 180°, which would equal one about 160°, according to him, east of the meridian of Greenwich, and is in fact, if we grant the above identification of the Gulf of Siam, approximate to 105° E. Africa he assumed to broaden indefinitely south of 10° S., and he further assumed a land connexion between south-eastern Africa and south-eastern Asia, so that he viewed the Indian Ocean as an enclosed sea. But in a review of the makers of geography such conceptions need be pursued no farther, save to observe that, as pointed out by C. R. Beazley, there was here a first approach to a realization of the bulk, though not the shape, of the land-surface of the earth. And Ptolemy's belief in a far eastern extension of Asia helped Christopher Columbus to believe both that he could reach Asia by sailing westward across the Atlantic, and, subsequently, that he had done so. The New World was elusive, and no word reached European geographers of the Chinese landing in North America about 500, if that took place, nor of that of the Norsemen in 1000.

Chapter VI

THE DARK AGE OF EARLY CHRISTIAN TEACHING

THE dark age of geography began before the Dark Age, so called, in history. Even before the decay of the Roman Empire set in, the advancement of classical science had ceased, and there supervened a period which yielded summaries of earlier work or (so far as geography is concerned) more or less otiose commentaries upon it, but nothing more. For long after Ptolemy's time there was no extension of the Roman Empire to add to geographical knowledge, and trade had settled into well-known routes so long as these remained open.

In the later part of the second and in the third century there was much disorder, including civil war, in the empire: the heart was weak, and no strength came from it to help the outer members against invaders—the Alamanni and Franks (A.D. 236), Goths (247), Persians (c. 260). Provincial emperors, so called, set themselves up, as in Gaul (259–69). At the end of the third century the empire was nominally restored; but trade was restricted, many cities and much good land were in decay, and culture declined as barbarian influences crept in.

About 324 a tremendous step, notably retrograde in respect of its effects upon geography in all departments except that of travel, was taken when, under the rule of Constantine the Great, Christianity was recognized as the religion of the Roman Empire. A new capital was established at Byzantium (Constantinople), and the division of the empire took place in 364. In this century the outer provinces were again deep in trouble: Britain was vacated by Roman troops in 407. In the first half of the fifth century the Vandals were making their conquests in north Africa. Rome was sacked, and the

fall of the Western empire followed in 476. Barbarian rulers
were now established in Italy, Gaul, and Spain, as well as in
Africa: new nationalities were in infancy, and knew nothing
of the old culture.

Of the historical developments of the next seven centuries,
certain few have principal bearing upon our subject. In 622
occurred the Hijra and the flight of Mohammed to Medina,
and the next hundred years witnessed the wide expansion of
Moslem power: Jerusalem fell in 637, Egypt was acquired in
640, in 711 the Visigothic ascendancy in Spain came to an
end, and the Arabs established themselves there and in
Sardinia, in 717–18 Constantinople was besieged but held
fast, and Christendom was saved. The cultural influence of
Islam was for long naturally spurned; but the Moslem advancement of learning was very well maintained in many
directions after Harun al-Rashid (*c.* 766–809) began to foster
the translation and study of Greek writers, and Moslem
geographical work, as will appear, was in its way remarkable.
In 787 the Northmen made their first descent upon England;
in 841 they appeared in France, and thereafter spread themselves farther afield over Europe; the Danes first wintered
and settled in England in 855, and in 878–9 Alfred of England
withstood them and made treaty with them. While they
added nothing constructive to the science of geography, they
disseminated knowledge of their own most interesting lands,
and of conditions of human environment which were (and
are) unique, and they inculcated a new spirit of adventure,
which is a spirit of use to geographers. The same spirit, as
also will appear, was present, but with a far different motive,
in the Christian missionaries, though of Christian cosmography no good can be said. We shall find that early Christian
teaching deliberately avoided pre-Christian geographical
theory, either holding like St. Ambrose (340–97) that 'to
consider the nature and position of the earth does not help
us in our hope of the life to come', or substituting its own

extravagances for the hated pagan views. The end of the world was widely anticipated in the year 1000; that expectation past, there was a strong religious revival leading to the Crusades (1095–1270), which further fostered the habit of travel. Meanwhile Arabic began in the same century to be recognized as the language of learning, and a start was made with the translation of Arabic scientific works into Latin. Whatever setback to geographical knowledge may be laid to the charge of Christian teaching for a thousand years and more, at least the spread of Christianity extended the known world, and the revival of learning, when at last it came, originated in the quiet of the cloister since there was no room for it to do so in the rough life without.

In the earlier part of the period under notice there appear certain notable examples of the Roman itineraries or distance tables worked out by careful measurement over the road-system. Such work was indeed Rome's best contribution to geography and had been going on from a much earlier date than that of the famous Itinerary of the Provinces of Antoninus Augustus; the results were certainly known to Polybius and Strabo and pretty certainly were used by Ptolemy. More than one emperor was called Antoninus: the particular emperor commemorated in the title of the Antonine Itinerary is usually supposed to be Caracalla (186–217); but the extant version is a revision attributed to the time of Diocletian (245–313). There is also attributed to the first half of the third century the so-called Peutinger Table named from Conrad Peutinger (1465–1547), a German antiquarian of Augsburg, who published a long-lost manuscript of it found by a contemporary scholar, Conrad Celtes, who made it over to him for that purpose. The table, certainly not the only one of its kind, showed a sort of diagram of routes by lines on a strip which exaggerated the distance of its west-to-east range (Britain to the mouth of the Ganges) as compared with that from north to south, but partook of the nature of a map in showing topographical features to some extent. Of the same century, but probably the second half, was the Stadiasmus of the Great Sea (the Mediterranean), the best example of the Periplus or collection of sailing directions, minutely detailed and clearly distinguishing the landmarks and accommodation for ships along the coasts. Lastly, the Christian pilgrimages to Jerusalem gave occasion for somewhat similar work, for in 333 the Jerusalem Itinerary (*Itinerarium a Burdigala Hierusalem usque*) was compiled for their guidance by a pilgrim from Bordeaux, immediately after the adoption of Christianity as the established religion of the empire. Subsequently the pilgrimages gave rise to a considerable literature down to the tenth century, but it is not of much geographical significance. An interesting survival, however, is found in the work of an unnamed student of Ravenna, commonly referred to as the Ravennese Geographer, who, probably about 650, drew upon a wide range of early authorities, irrespective of creed, in compiling a kind of gazetteer of the world. He certainly used the material of Roman itineraries and also of Ptolemy (whom in his ignorance he called the king of the Macedonians in Egypt), and he referred to a large number of other authorities, which suggests how much more extensive than we know was the geographical or quasi-geographical literature down to his time. The Ravennese Geographer's results are not good: his inaccuracies as a compiler are many and to originality he has no claim.

It may be said broadly that in this period there are plenty of writers about geography but no geographers. We need do no more than cite a few outstanding examples illustrative of various outlooks.

There has always been a popular phase of geography—in our own time even Louis de Rougemont was not the latest purveyor of it—which has associated the marvellous with the unknown. With deference we might cite the Homeric poems as an early repository of such writing; the taste for it persisted through the classical period, and became strong in the Dark and Middle Ages. The want was met (and not about unknown lands only) by Gaius Julius Solinus, who wrote about the year 250, and became known as Polyhistor from a title, which might be loosely translated as snippets, later given to his work *Collectanea Rerum Mirabilium*, the collection of wonderful matters. He has been described as a bombastic plagiarist of Pliny and Pomponius Mela, and we may leave him at that, noticing him only as an example of a type, and as a writer who through the popularity of his work had a very bad influence upon Christian geography long after his own time.

Of a different stamp was the last important Roman historian, Ammianus Marcellinus, who was born about 330 and died not earlier than 391. Though he wrote at Rome and in Latin he was a Greek of Antioch; he had military training and experience, and he recognized the value of geography in relation to history. He made free use of Ptolemy, and the Greek tradition in historical geography reappeared in his work.

Cosmographical questions became deeply involved. While the doctrine of the spherical earth had come to be accepted, in the period preceding the Christian era, by educated persons who paid any attention to the question, it is not to be supposed that it had achieved anything like its common acceptance to-day. The flat earth survived as a popular conception. Thus Lactantius Firmianus (*c.* 260–340), who became converted and wrote his *Divinarum Institutionum Libri Septem* in middle age, denied the conception of a spherical earth and its implications (such as the antipodes), not only as being against the teaching of the Bible but because he personally thought them impossible. On the other hand, Martianus Capella, of the early part of the fifth century, who lived apparently in Carthage and was not converted, in his encyclopaedic work *Satyricon* still maintained the sun to be the centre of our system. These views may be cited as opposite extremes for the period. Paulus Orosius, who flourished about 415 and as a disciple of St. Augustine wrote the *Historiae adversum Paganos*, prefaced his work with a summary sketch of the known world which in some respects seems to throw back curiously far into pagan chorographical authority—it was of necessity only pagan cosmography that was held in abhorrence. Orosius placed the earth in the centre of the oceans, had nothing of Ptolemy's conception of its extent, and incidentally maintained the supposition that the Caspian Sea communicated with the ocean to the north. He was the first writer known to us to distinguish Asia Minor from the major part of the continent.

Christian cosmography reached an extraordinary development in the work of Cosmas of Alexandria in the sixth century. He is to be regarded as an example of the extreme; not as a writer whose influence lasted long or spread widely. He was a maker of geography indeed, but it was a geography of his own. He was a merchant in earlier life, and travelled fairly widely, as far south as Abyssinia, and into the Indian Ocean, probably visiting the Persian Gulf, the west coast of India, and Ceylon. He afterwards became a monk and set down an account of the lands he had visited. To his reputation as a cosmographer the name by which he is known to us is very likely due, as the surname of Indicopleustes is intended to commemorate his eastern voyaging. After his first essay in geographical work he composed his Christian Topography, apparently in the period 535–47. His central object was the refutation of pre-Christian cosmography: he was not content to ignore this, or to say that the figure of the earth was no part of Christian teaching. He therefore worked out an earth modelled in all detail upon Moses' tabernacle. This earth was flat, rectangular, and oblong, twice as long west-and-east as north-and-south, and surrounded by oceans. A high mountain rose in the north, and the sun, quite small and near the earth, revolved around it, being hidden behind it at night. Cosmas conceived lands beyond the oceans, where was Paradise. The arch of heaven was affixed to the earth along its edges, and the whole rested upon the stability of God. Cosmas had read widely in order to overwhelm by argument; and Aristotle, Eudoxus, Ptolemy, Pytheas, and

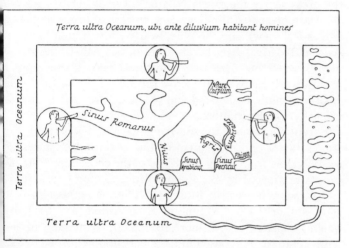

FIG. 6. The World according to Cosmas Indicopleustes, *c.* 540 A.D.

other names familiar to us come under his ban. It is probable that the crude diagrams and illustrations which accompany the earliest existing manuscript of the Christian Topography (10th century) are Cosmas' own; if so, they are the earliest extant examples of Christian cartography.

This did not reach a high standard, save perhaps that some of it should be accorded a certain artistic merit. The known world was shown either as a rectangular oblong, or as a circle, or oval, but not with any idea of projection—these figures represented different ideas of its actual shape. The maps generally were centred upon the holy city of Jerusalem, and the inhabited world was surrounded by ocean, of which the

Caspian Sea was usually a gulf. The maps were devoid of any topographical accuracy or attention to scale. One class of map was so far diagrammatized as to take the form known as T-O. Here the O represented the ocean, and the T was within it, the upright representing the Mediterranean Sea, the cross-stroke the rivers Nile and Tanais (Don). As the Tanais represented the boundary between Europe and Asia (a classical view), and the Nile, nearly enough, that between Asia and Africa, Asia became the upper half-circle of the map, Europe and Africa the lower quarters. There appears in these early maps the attractive practice of portraying living creatures, both of the land and of the sea, as well as appropriate verbal descriptions.

Exceptional among the surviving maps of this period is the Anglo-Saxon or Cotton map of the world, which takes its second name from being preserved in the collection of manuscripts originated by Sir Robert Bruce Cotton (1571–1631) and now in the British Museum. The map is assigned by some authorities to the time of King Alfred (871–901); by others, apparently with more likelihood, to the end of the tenth century. It is based in large part upon Orosius, in less degree upon earlier writers back to Pomponius Mela, and in some measure upon knowledge acquired as a result of the incursions of the Norsemen in the two or three centuries preceding the compiler's own time. There is some evidence in favour of its compilation by an Irish scholar, and in this connexion we may refer forward to the wanderings of Irish monks in the far north, recorded by Dicuil, and to be mentioned later. The map is associated with a manuscript of Priscian's *Periegesis* of the fifth century, but this did not afford the cartographer any independent material. The map was designed to illustrate, among other matters, the distribution of the twelve tribes of Israel, but, this apart, it reveals geographical scholarship of an order much above other maps of its period.

Some of the earliest Christian maps showed the unknown habitable land—the south temperate zone—beyond the hot uninhabitable equatorial belt. This belief, of classical origin, became general: it was only in relation to the sphericity of the

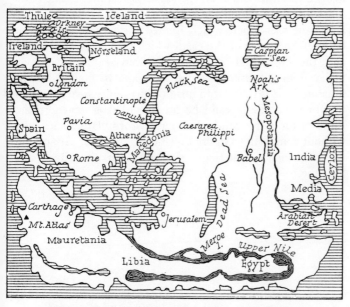

FIG. 7. Outline of Anglo-Saxon Map of the World.
(Names modernized. In the original the East—India, Ceylon, &c.—
is at the top.)

earth and its position in the universe that early Christian teaching generally denied or held aloof from classical theory. And to that teaching there were quite early exceptions. There is a distinct suggestion that the higher intellects among the Christian fathers shunned the subject partly because their own understanding would have bidden them adopt the pre-Christian view, and that in teachers of the school of violent denial (like Lactantius and Cosmas), for all their diligence in reading, we have the type of ill-educated bigot. Whatever the view of the Venerable Bede (*c.* 672–735) as to the figure of the earth, he clearly did not believe it to be a flat disk. In its relation to its surroundings of water, air, and fire, he likened it to the yolk of an egg (a classical simile). The outermost fire from which the sun derived and transmitted heat and light was a belt which limited life to temperate zones between the poles of the earth which received too little heat and the equatorial zone which received too much. He believed in a southern temperate zone, habitable, but not inhabited. It is difficult to imagine any figure save the sphere to which his ideas could be fitted, and it is ungenerous to think of the gifted student of Jarrow formulating such ideas without conceiving a figure to which to fit them. At any rate, the doctrine of sphericity gradually reinstated itself after his time.

Plate II. Portion of map of Egypt, Sinai, Palestine, Syria, and part of
Mesopotamia, done in mosaic on the pavement of a church at Madaba
(Medeba, NE. of the Dead Sea). Probably of the first half of the sixth
century A.D. The towns, shown by pictures of buildings, are largely
places of pilgrimage. A part of the Nile delta is seen in this photograph.
The map is lettered in Greek

Chapter VII

MOSLEM GEOGRAPHY

THE Moslem culture to which reference has already been made is of peculiar interest in relation to geography. Islam had no natural centre for a single empire from which political accretions to the north, east, and west could be controlled. Under Walid I, in 705 and following years, Moslem armies were operating as far east as the Indus, Moslem fleets as far west as Sicily and Sardinia. But as they spread their faith among other peoples it was in the natural course that the Arab should be out-Moslemized by his converts, and the rise of the Abbasid dynasty in the East, based upon the claims of Persian followers of Islam, provides an example. Similarly in the west other dynasties not of Arab origin achieved temporary power. It has been said that by the middle of the eighth century the new power was Moslem, not Arab, and this statement may be exemplified by another: that out of some sixteen geographers of note from the ninth to the thirteenth century four were natives of Persia, four of Baghdad, and four of Spain. In Spain, before the coming of the Arabs, there were many Jews, descendants of deportees or emigrants from Palestine, who preserved the traditions of Alexandrian learning. But the Arab must have possessed a strong fundamental geographical instinct, born of subjection, in his original home, to stringent conditions of geographical environment. A curious illustration of that instinct is found in Spain, where after the first Arab invasion in 710, Abdul Rahman, a survivor of the overthrow of the Ommeyad dynasty by the Abbasids in the east, arrived in 756 and succeeded in establishing a dynasty, independent of the eastern caliphate, which lasted until 1031. Here at the farthest western limit of his migration the Arab carried with him a remembrance of the East, and labelled sites and

E

features of the peninsula according to supposed resemblance
to those which he had known, so that there appeared in Spain
a new Damascus, a Jordan, even a Palestine and an Egypt.

Both the administrative and the commercial activities of

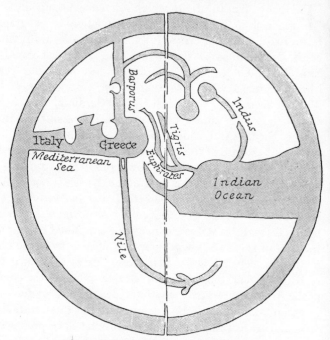

FIG. 8. The World according to Ibn Haukal.
(In the original the south is at the top.)

the widespread Moslem states invited the study of geography.
Thus Ibn Khurdadbih (of the 9th century), was a chief post-
master. Istakhri and Ibn Haukal, who wrote in 951 and 978,
were travelling merchants. In spite, or because, of the scattered distribution of centres of population and trade there
was constant movement of travellers between one territory
and another, and it is recorded by Ibn Haukal that at Tarsus,

one of the Syrian frontier strongholds, every important city from Persia to north-west Africa maintained a rest-house for its visiting citizens.

Following Harun al-Rashid, the caliph Al-Mamun of Baghdad (died 833) fostered astronomical and geographical work by having Marinus', Ptolemy's, and other works translated. The circumference of the earth was recalculated, and degrees were measured. A calculation was made, for example, which gave an error of only three degrees between Toledo in Spain and Baghdad, and the Moslem measurement of the length of the Mediterranean Sea from west to east was nearer the truth than any classical estimate. Moslem cartography, however, seems never to have reached a standard at all commensurate with geographical writing. The Arabs introduced from the east, and perfected, the planispheric astrolabe, which was used for reading astronomical altitudes and calculating latitudes, time, heights of mountains, &c., and from it was adapted in the fifteenth century the navigators' astrolabe as used by Columbus. This was not superseded until toward the middle of the eighteenth century.

Works of the gazetteer type, and descriptions of travel, became very numerous. The description of the world by Mohammed ben Musa furnished a list of names of places and positions by latitude and longitude. Ibn Khurdadbih's work consisted mainly of a list of routes and distances, an administrative survey, but it included a cosmographical survey which made use of the comparison of the earth to the yolk of an egg, and reproduced the view of the equatorial zone as desert by reason of the heat, and of the uninhabited though habitable antipodes. His descriptive matter carries us definitely as far as China, Japan, and Korea. Like so many other Moslem writers he was something of a fabulist. But we should not omit to recall, even in a review of the work of makers of geography, that it is to this trait that we owe the famous legend of Sindbad the Sailor, which is based upon narratives, pretty certainly true, of Moslem travels in the ninth and tenth centuries, though overlaid with Greek, Persian, and Indian tradition.

In the first half of the tenth century Al-Masudi of Baghdad (died 956) visited India and Ceylon, probably China, Madagascar, the Caspian region, Syria, and Egypt. In his work, *Meadows of Gold and Mines of Precious Stones*, he applied the results of travel and personal observation to history; he was commonly compared with Pliny. The Moslem demand for geographical works at this period appears to have been great, from the fact that Abu Zaid's work, written about 921, was revised thirty years later by Istakhri, who travelled all through the Mohammedan lands, in his *Book of Climates* (or zones); and this was again revised and extended by Ibn Haukal in 977 in his *Book of Roads and Kingdoms*. To this period belongs Avicenna (980–1037), a native of Bokhara, the student of Aristotle whose *Canon of Medicine* survived as a text-book in certain French universities down to the middle of the seventeenth century. He is credited (not certainly) with ideas as to the elevation of mountains by folding, their sculpturing by erosion, and also the long period necessary to such processes, which appear almost to anticipate modern physiography. Orthodox Christian beliefs concerning the creation, as will appear, were to prohibit the adoption of such ideas in the West for several centuries yet.

Idrisi (*c.* 1099–1154), born in Spain and probably educated at the great centre of learning, Cordova, travelled in north Africa and Asia Minor. He afterwards lived in Sicily, and there made a celestial sphere and a map of the world in silver for King Roger II. Idrisi compiled to the king's order a description of the world based upon the reports of observers who were sent out by Roger. Political and natural divisions were not used as the basis for this work; instead, Idrisi divided the known world into seven latitudinal climates or zones between the equator and the far north, and each zone into eleven parts by lines at right angles to those of the latitudes.

In order to complete the story of Moslem geography for our purpose we may depart from chronological order and refer to Ibn Batuta of Tangier (1304–78), who during thirty years is estimated to have travelled 75,000 miles. East Africa, India, and the Malay Archipelago, Syria, Arabia, and Asia Minor, the regions of the Black and Caspian Seas were all known to this wanderer, and his descriptions were able and accurate; but they did not become known to Europe for some centuries after his death.

FIG. 9. The World according to Idrisi.

(In the original the south is at the top.)

Moslem science was declining by this time, but reference is also due here to the work of Ibn Khaldun of Tunis (1332–1406). He has been styled the greatest historical thinker of Islam, and he made a remarkable application of geographical considerations to the history of his people. Throughout most of the lands over which Islam had extended its sway, as well as in its original Arabian home, arid tracts are interspersed with areas capable of cultivation. Ibn Khaldun saw the fundamental distinction between the inhabitants of lands of these respective characters; those of the arid country are nomadic, those of the cultivable country, settled. He envisaged the nomad life as connoting no high standard of culture, but as inculcating the tribal instinct, readiness for warfare, and bravery. He placed the nomad not only lower in the cultural scale but also earlier than the settler of the agricultural lands, and he noted the civilizing effects of contact with the settler upon the nomad. He observed the adverse effects of civilization and luxury upon the primitive virtues of the nomad, and the sequence from the zeal of the primitive conqueror through the power of the established empire, to the decay of that power, which is visible (and was so in his own time) in the history of Islam. He argued from this particular case to human history in general.

Chapter VIII

THE BEGINNING OF THE RENAISSANCE IN EUROPEAN GEOGRAPHY

AS the rise of the Moslem culture overlapped the dark age of the Christian, so did its decline overlap the beginning of the Renaissance in Europe. Before dealing with that, it should be attempted to review the growth of knowledge of the world which helped to give rebirth to geography in Christendom.

The trade routes of the Roman empire were maintained long after the fall of the empire—some of them have their modern counterparts still. Naturally the main lines, which alone concern us, ran west and east, for within the empire itself the Mediterranean was the great seaway, and the Orient supplied the bulk of luxuries not obtainable within the Mediterranean area. Only such limited traffic as that in Baltic amber on the one hand, and gold and ivory from the east African coast lands on the other, connote long trade routes north and south. In the latitudinal direction we find a route from Europe by Byzantium, Trapezus (Trebizond), the Caspian Sea, and the Oxus River to Samarkand, where contact was made with trade to and from China in one direction, India in another. A second route followed the Euphrates and the Persian Gulf coast to India. A third struck eastward from the head of the Tigris-Euphrates plain through Persia and thence north-east to the Oxus, joining the first. Along such lines passed a gradually increasing knowledge of India and China, together with a fine store of romance. Silk from China had become an important article of the luxury trade in Rome before the beginning of the Christian era. A famous story tells of two Persian monks who, having lived in China and learned the methods of rearing silkworms, brought back eggs in a hollow cane to

Justinian at Byzantium, so that then and not till then were the secrets of the production revealed to the Mediterranean world. Embassies to China took place as early as 166 and 284. Christian missionaries first reached India in 189 and had penetrated far into Ethiopia before the Roman empire had accepted Christianity (324): they reached Abyssinia about 330. About the same time missions were pushing into central Asia, and a Nestorian mission, which afterwards flourished widely for a while, settled in China in the seventh century. Reference has been made already to the practice of pilgrimage to the Holy Land which spread all through western Europe, including the Scandinavian lands as they became Christianized. The Jews had also their pilgrimages, and travelled much as traders and emissaries. The Rabbi Benjamin of Tudela in Navarre travelled through Egypt, Assyria, Persia, and central Asia to the confines of China. He wrote of his wanderings of thirteen years (c. 1160–73) in an itinerary, which is a valuable record of Jewish activity at the time; but Christian hostility caused the neglect of his work.

As for the far north and north-west, the Irish monk Dicuil, who completed his work *De Mensura Orbis Terrae* in 825, recorded that in 795 Irish hermits had visited 'Thule', remaining there from February to August, and returned marvelling at the nightless midsummer when there was 'no darkness to hinder one from doing what one would'. Without doubt they had been in Iceland; and early in the year they even went a day's sail to the north of it, when they encountered ice-floes. In regard to his work generally Dicuil was widely read in the classics as well as in later writers, and he made use of surveys of the Roman empire which are not known to us. As we have seen, it was not long after his time that the Scandinavians' migrations overseas brought them into intimate contact with western Europe; and Adam of Bremen, a canon of the cathedral there and a geographer enlightened as to the sphericity of the earth, made in his

History of the Church of Hamburg (*c.* 1075) a close study not only of north German and Baltic, but also of Scandinavian lands including the Norse colonies westward across the North Atlantic Ocean. Adam made the first surviving mention of Vinland, that unknown shore beyond Greenland, which after long dispute appears to be reasonably assigned to the North American coast between Nova Scotia and Massachusetts. But European geography could not rise, on such slender evidence, to the conception of a new continent.

Before we leave this period, reference is due to the East Roman emperor Constantine VII, called Porphyrogenitus (905–59), a writer and artist himself, who also inspired much writing and art in others. His use of geography was in the spirit of Strabo: his *De administrando Imperio* gave an account of existing conditions both in the territories of the empire and in those of some of the bordering lands to the north, east, and west; and this is of real value.

This sketch may serve us down to the twelfth century, and it seems that we are to imagine a world in which a good deal of movement was going on, and in which opportunity was afforded for contact between Spain and China, between Scandinavia and Arabia; but that contact had not yet brought accurate knowledge of the geography of the outer fringes. Nor, indeed, did it do so for many years to come, and this is intelligible when we remember that the printing of books, as we understand it, did not come into practice in Europe until the fifteenth century, and that the scribes of earlier date had enough to do in transcribing known books without going out to seek what was new.

Nevertheless, in Europe of the twelfth century something approaching a revolution took place, not only political but also intellectual, largely on account of the Crusades. Directly out of these grew an increase of commerce and a spread of individual knowledge of the world. Feudalism, and a certain parochial outlook, began to be broken down in western

Europe; the power of thrones increased, and the foundations were laid of that nationalism which became apparent
in the two following centuries, and created the spirit of
rivalry which had so much to do with the extension of
exploration. France became a powerful centralized monarchy.
Spain gradually recovered the peninsula from the Arabs. As
regards its own political history, there falls within this period
(from *c.* 1100) the growth of Portugal into an independent
kingdom; and from Portugal and Spain, we shall find, much
of the inspiration of overseas exploration was to spring. The
passage of crusading traffic through the north Italian ports
led to great increase in their wealth and power. A like development in Germany was only delayed by the struggle
between the Holy Roman Empire and the papacy.

The power of the Church also was enhanced along with
the growth of the crusading spirit. Learning was still inspired in and from the Church almost exclusively. In the
twelfth century many new monastic orders were founded.
Scholasticism had originated with the attempts of Charle-
magne (*regn.* 768–814) to foster education by bringing learned
men to his court and establishing schools in the major
religious foundations: it was a result of this development that
the distaste of Christian for pre-Christian philosophy began
to weaken. The great age of scholasticism falls within the
thirteenth century, the period of Albertus Magnus and
Thomas Aquinas. It is to be noted, moreover, that the names
of English scholars take a prominent place in this part of our
survey.

Thus, Adelard of Bath, the scholastic philosopher, who
died in or about 1187, travelled in western and southern
Europe, north Africa, and Asia Minor. He became specially
interested in Arabic works, and among other translations he
made available for European use the ninth-century Arab
tables of latitudes and longitudes. He also wrote a treatise on
the astrolabe. His tables were supplemented, among others,

by a series worked out by Moslem and Jewish computers under the instruction of Alphonso X of Castile (1252–84), which yielded a reckoning of the length of the Mediterranean axis much nearer the truth than Ptolemy's.

But English science has been said to 'begin' with Roger Bacon (*c.* 1214–94), who after studying at Oxford proceeded to Paris (where the university had been founded in 1215), and acquired early a wide acquaintance with Arabic authorities. Returning to Oxford, he entered the Franciscan order probably about 1250. His geographical work appears in the fourth part of his *Opus Majus*. One of his principal claims to greatness lies in his insistence upon the importance of experiment in scientific work, and he was an opponent of the scholastic theologians in so far as they withheld from scientific method. In geography, however, this attitude did not differentiate his work so greatly as in other subjects from that of others of his time and before. It is true that he urged that accurate mapping could only be based on accurate determinations of latitude and longitude. Hipparchus had said that, but it needed repeating. It was otherwise characteristic of the time that earlier results were accepted without question as basis for argument. Bacon took the length of a degree at the equator as $56\frac{2}{3}$ miles (nearly) and obtained an estimate of the circumference of the earth only one-fourteenth less than the truth. But in accepting the sphericity of the earth he was doing no more than was done by other great Christian scholars of the time. Nor did he doubt the current Christian doctrine of the earth as the centre of the universe. Nor, again, did his regard for the proof of facts by experiment, or experience, prevent him from theorizing as to the distribution of land and sea. At the outset of his geographical work he divided the earth, for the purpose of laying down this theory, into four quarters by means of the equator and a great circle passing through the poles. He assumed the inhabited world of the northern hemisphere to be probably balanced by corresponding southern land-masses, and he took the northern lands to extend much more than half way round the globe: that is to say, he believed in a continuous land-mass extending eastward from the North Atlantic Ocean much farther than 180°. He was not alone in this belief, which in a modified form dated from Ptolemy; but Bacon and others of his period emphasized the error, which persisted until the fifteenth century, and Bacon's own work, among others, is known to have influenced Columbus in seeking the Indies by way of the western ocean.

Albertus Magnus or Albert of Cologne (c. 1206–80) was primarily a student of Aristotle, and in his and other work of the period there is found a revival of interest in physical geography. Thus, it now began to be appreciated in medieval scholarship that climatic conditions are not determined rigidly according to the ancient zones, but are affected also by height and exposure of the land, and that these conditions of environment have effects upon vegetation and living creatures. Ideas are also encountered about the hot centre and cooling crust of the earth, about the folding of the crust, about earthquakes, and erosion and deposition by water. We must not, however, attach too much value to these conceptions at this period. Little real progress was made on this side of geography. Christian dogma, which assigned to the third day of the creation the separation of the land from the sea, practically prohibited any advance in the study of that division of our subject which impinges upon geology. Catastrophic episodes such as the flood and volcanic eruptions had to be invoked to account for the visible evidences of changes upon the earth's surface, in addition to such geological phenomena as the occurrence of marine fossil forms high above the sea-level. Even the scientific genius of Leonardo da Vinci two centuries later (1452–1519), for whom travel-narrative and cartography had a strong attraction, could not in this direction win contemporary thought from the paths of orthodoxy.

Cartography, so far as concerned maps of the world, was necessarily of slow development. There can have been only

Fig. 10. The Hereford Map of the World (simplified.)
(In the original the east—India, &c., with Paradise—is at the top.)

a very narrow circulation of maps. At this period we find two main groups which may be called churchmen's and sailors' maps, showing little evidence of common knowledge, and none of method, between one class and the other. The world maps identified with various religious foundations, as exemplified here in the Hereford map of about 1280, merely elaborate, at the best, much earlier work. The general principles are similar. There is no evidence of acquaintance with the mathematical or physical sides of geography (although, as we have seen, ecclesiastical scholars like Bacon were acquainted with both, and Bacon's own work was illustrated with diagrams of the *climata* or zones of the earth). The pictorial attributes of monastic cartography—the sketches of towns, animals, strange men, and so forth—were maintained; so were the descriptive and other notes on the face of maps. Intermediately between these and the sailors' maps comes such work as Idrisi's, on a scientific basis according to his lights, but of no great merit.

The sailors' maps of the Mediterranean Sea, commonly known as *portolano* maps because they accompanied the *Portolani* or sailing directions (Ital., *porto*, harbour), were of a different character and of a much higher standard so far as concerned the well-known coasts. They are first mentioned in 1270 (in connexion with a crusade), and the earliest extant example dates from about 1300. But they are taken to have developed gradually from earlier times, just as our own current Admiralty charts in some instances are revisions based upon originals of quite respectable antiquity. The material for the eastern Mediterranean area has been dated by some authorities back to Byzantine or even earlier sources, but without direct evidence; for the west, it appears that in the compilation and improvement of these maps (which have been called 'the first true maps') the seamen of northern Italy and Catalonia contributed most largely. E. G. Ravenstein [1] thus describes the *portolano* maps: 'These charts are based upon estimated bearings and distances between the principal ports or capes, the intervening coast line being filled in from

[1] Ravenstein's historical article under the heading *Map* in the *Encyclopaedia Britannica*, 11th ed., is especially noteworthy for his own series of sketches illustrating more clearly than would have been possible by direct reproduction the main features of a number of the old maps.

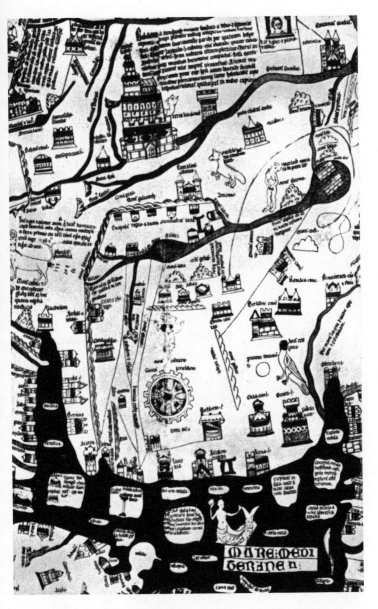

Plate III. Portion of the Hereford Map. Jerusalem is shown in the centre, the Tower of Babel in the upper part of the illustration

Photograph, W. H. Bustin

more detailed surveys. The bearings were dependent upon the seaman's observations of the heavens, for these charts were in use long before the compass had been introduced on board ship.'

The compass is first mentioned in European literature

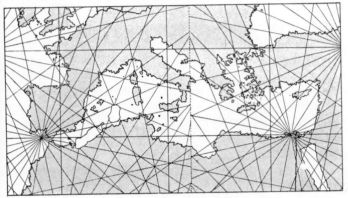

FIG. 11. Portolano chart of the Mediterranean.
(Juan de la Cosa, 1500.)

by Alexander Neckam of St. Albans (1157–1217), but not, apparently, as a novelty in his time: clearly it came gradually into use on ships during the period under review. Its use must have helped to improve the *portolano* charts, but only slowly. As they stand at this early period, however, they are remarkably accurate in the details of the Mediterranean coasts, and also in the length of the sea. Some are provided with a scale of miles known as *portolano* miles, the origin of which is uncertain; they coincide closely with a Catalan measure, but have also been taken as of Greek or eastern origin. In orientation the maps are less correct, and this has been attributed to errors made when the cartographers first attempted to fit together maps of parts of the sea into a whole. Similarly, it was not always realized that the *portolano* miles were shorter than the Roman miles in use apart from these special maps, so that errors supervened when the maps (as we shall see) were extended beyond the Mediterranean area.

The maps, even when, as in some instances, they were provided with a network, did not show meridians and parallels. They were commonly covered with rhumb-lines radiating from a series of centres, no doubt to assist sailors in following their courses, though the maps are obscured by these lines, which often bore the initials of the names of the principal winds.

In the *portolano* maps it is possible to recognize the genesis of a school, or series of schools, of cartography, for by the fourteenth century much work was being done, and the names of a considerable number of cartographers are known, in Genoa, Venice, and other Italian ports, in Majorca and in Catalonia. To extend the maps to cover the known world was a natural development, but before dealing with this extension we shall examine the new material provided by travellers, traders, and scholars, in order to be able to judge how far the cartographers used it.

Chapter IX

THE MISSIONERS TO THE EAST

THE principal incident in the political history of the thirteenth and fourteenth centuries, connected with the history of geography, is the rise and expansion of the Mongol power. Jenghiz Khan, during his reign of conquest (1206–27), acquired a dominion which extended from the Far East to the river Dnieper. In 1235–6 the Mongols were overrunning Mesopotamia, Armenia, and other territories south of the Caucasus, and in 1241 they were ravaging Hungary and Poland. In 1258 Baghdad fell to them, and from 1259 to 1294 they reached the height of their power under Kublai Khan. That power lasted not long. The fall of the dynasty of Jenghiz was completed by 1368; the Mongols, apt at conquest, were the reverse when it came to consolidating their gains and imposing their rule upon defeated peoples. Yet they were not unready to acquire culture: thus (as appropriate to our subject) we find that Hulagu, in Baghdad, furthered the establishment of an astronomical observatory, which contained among other geographical equipment a fine terrestrial globe. Nor were they averse from receiving emissaries from the Christian powers of the west; and so it came about that a great accretion of knowledge of Asia was conveyed to Europe in the thirteenth century.

In 1245 Pope Innocent IV sent a Catholic mission under Joannes de Plano Carpini, of Umbria, who travelled through Poland, Russia, and Turkistan to find Kuyuk Khan near Karakoram, and returned in 1247. In that year another mission was sent by the Pope to the Mongol Batu in Armenia. In 1253–5 William of Rubruquis, a Franciscan, carried out a visit to the Khan Mangu at Karakoram at the behest of Louis IX of France, reaching south Russia by crossing the Black Sea, and thence riding eastward over the steppes.

There were other legations in the field about this period. Hayton, King of Little Armenia, obeyed a summons to wait upon his overlord in central Asia; and on the outward part of the journey, which occupied most of his period of absence (1253–5), had to travel through enemy (Moslem) territory in disguise. But he returned with speed, under Mongol escort, along a route which was new to Western knowledge, by Ili valley, Kulja, the Syr-darya, Samarkand, Merv, and northern Persia. The reports of Carpini and Rubruquis, however, were naturally the more valuable to western Europe: that of Rubruquis especially, by what was in those times a fortunate chance, was put to almost immediate scholarly use, for Roger Bacon quoted him freely in his *Opus Majus*.

About the middle of the thirteenth century there were resident in Constantinople brothers of a well-born Italian family, Nicolo and Maffeo Polo, merchants. A trading venture took them to the Crimea, and thence to Kazan and Bokhara, where they fell in with a party of Kublai's Khan's men, who, no doubt instinct with the Khan's desire for contact with Western culture, carried them to his court in China. He received them well, and sent them back with a message to Rome, asking for a papal embassy to visit him. During their long absence Nicolo's wife had died, but he found awaiting him a son of fifteen years, whose name was to become famous. Marco Polo (*c.* 1254–1324) accompanied his father and his uncle to China when they returned thither in 1271; and none of them saw their home land again for twenty-four years. During that time Marco Polo rose high in the service of Kublai Khan. His journey outward had been by way of the Pamirs and the central Asian desert; during the years of his service in China he had occasion to travel widely, and his return home was made by sea, by way of Sumatra and India, and Persia. His book was dictated in 1298–9 during captivity at Genoa, whither he was taken after a naval battle between Venetians and Genoese off Curzola in the Adriatic Sea.

No preceding traveller compared with Polo in wide knowledge of Asia. In central Asia—the parts about Lop-nor—he described ground not again explored by a European till 1871, and within the compass of his book he covered not only China in detail, but also the surrounding lands, including the whole of eastern Asia from the Arctic territory to the Malay Archipelago. He had much to tell of Ceylon and India, and he was well informed about Abyssinia, and to some extent about the east coast of central Africa. As a geographical writer Polo stands very much higher than a narrator merely of his own travels.

The Christian missions continued to provide important sources for the descriptive geography of eastern Europe, central Asia, China, the Malay coast, India and Ceylon, Persia, and south-western Asia. Examples of their work are provided by Joannes de Monte Corvino in Italy, who about 1290 visited Persia, India, and China; by Jordanus of Severac in France, whose work is chiefly concerned with India (c. 1321); and most notably by Odoric of Pordenone. He travelled by way of India and Malaya, and about 1323–8 was at Peking, where the Franciscans had their principal church among a number in China. Odoric returned to Europe through central Asia. He was a genuine travel-geographer as well as a missioner. Of such there are others whose work survives, and there must have been many whose knowledge was passed on only orally. But of world-geography the missioners displayed little knowledge, if any; indeed, when such a one as Joannes Marignolli of Florence (c. 1338) embarked upon geographical theory as associated with the earthly paradise and other semi-sacred conceptions, his imagination was only less ill-directed than that of Cosmas.

At this period Venice, Genoa, Pisa and Florence, Marseilles and other French cities, Catalonia and Aragon, England, Germany, and the central Russian trading centres were all interested, more or less competitively, in trade with distant (that is mainly Eastern) lands. It is not without geographical significance that the origin of the Hanseatic League of trading centres is traced in Germany in the second half of the thirteenth century. Geographical literature which accompanied the development of commerce is not wanting. Marino Sanuto of Venice, about 1313, curiously combined religion and commerce, and Francesco Balducci Pegalotti of Florence produced what may be called a typical commercial handbook early in the fourteenth century. Of yet another type was the Bavarian Schiltberger, who, whatever his defects in execution, deserved merit for attempting to pass on the knowledge he acquired, over a range from Siberia to Arabia, during more than thirty years' captivity with the Turks and Tartars, which terminated in 1427.

Thus advancement in the knowledge acquired by travel took place during this period mainly overland. But at the end of the thirteenth and early in the fourteenth century Mediterranean seamen, principally Italian, were passing westward and southward into the ocean beyond the Straits of Gibraltar; and Madeira, the Azores, and the Canary Islands began to be brought back to the knowledge of men. C. Raymond Beazley associates this movement with the fall of Acre in 1291, when the Mameluke Khalil captured the city from the 'Franks', after which the last of the Crusaders were soon driven from the Holy Land. These events may well have turned the minds of some of the traders to the Levant and beyond toward the possibility of a sea-route to their more distant objectives, such as was to be achieved in the course of the fifteenth century.

No outstanding factor in our history, however, emerges during the fourteenth century. Nor was there yet any centripetal force in geography as a science. No doubt it was an attempt to provide one which led to the translation of Ptolemy's geographical work, begun by Emanuel Chrysoloras, a Byzantine teacher resident in Italy, and completed in 1410 by his pupil Jacobus Angelus. Both text and maps were done into Latin, and the work met a need and in a measure fulfilled its function, though its results were not wholly beneficial. For example, the Medicean *portolano* map of 1356, and the Catalan atlas of 1375, gave a right idea of the peninsular part of India, for which evidence certainly existed both from sailing directions for the Indian Ocean and from the works of the missioners. Yet Fra Mauro's map of 1457 and Martin Behaim's globe as late as 1492 reverted to the Ptolemaic conception, showing no peninsula, but an island immensely exaggerated in comparison with the actual size of Ceylon. The Atlantic islands, the rediscovery of which has been mentioned above, appeared in *portolano* maps about the middle of the fourteenth century: the Medicean map has the Azores and Madeira for the first time, so far as is known, under their modern names. It was natural that this should be so; on the other hand, there is less evidence than might at first thought have been expected for the appreciation of the work of the missioners and other land travellers by the cartographers and compilers. Sanuto appears to have known nothing of Marco Polo, and the geographical parodist Jean de Mandeville or Sir John Mandeville, of the period 1360–70, knew him hardly at all, though he made ample use of Odoric and others more nearly his contemporaries. Nevertheless the Catalan atlas, a remarkable work for its period, based its representations of India, central Asia, and the Far East upon Polo's work. There is a suggestion here that a term of half a century or more was necessary for the knowledge yielded by travel to reach and be used by the map-makers of the time. It is not easy for us to realize the slowness of the dissemination of knowledge which is still characteristic of the history of science generally at the period under review; but it is fair to geographers to recognize this as one of the major difficulties under which they worked.

Equally difficult it is for us to recapture the sense of

romance which would naturally accompany a period of intellectual life in a world still so little known. It has not fallen within our scope closely to examine earlier geographical myths. But the period of the thirteenth and fourteenth centuries was one of preparation for that period in which the foundations of modern geography were powerfully strengthened. Therefore reference is due here, by way of illustration of the spirit of the time and the struggle of fact against fiction, to a particular fable, unique as in some respects so explicit, and so widely believed. Before the middle of the twelfth century there first arose in Europe the belief in a Christian king in India or the Far East, to whom became attached the style of Prester John. A bombastic letter purporting to be addressed by this potentate to the Byzantine Emperor Manuel Comnenus about 1165 was widely circulated and caught the public interest—not only on romantic but on political grounds, for in this Asiatic power was seen a possible ally against the Moslem. Modern criticism has associated the fable in successive versions with various Eastern chieftains including Jenghiz Khan, and so, indeed, did the missioners in Asia to whom we have referred. Some of these visited the country north-west of Peking where a prince to whom the name of Prester John, or at least descent from him, was assigned; but Odoric of Pordenone showed that the stories of its richness and the splendour of its courts were wildly exaggerated, and Prester John as a figure in contemporary Asiatic mythology now faded away. But he made a successful reappearance in Africa. By the early part of the fourteenth century he was located (while the Asiatic story was still current) in Ethiopia, and then more definitely in Abyssinia (where Christianity had been introduced as early as 330), and this allocation, as will appear, was still explicit at the close of the fifteenth century in the practical minds of the Portuguese explorers.

Chapter X

PORTUGAL OVERSEAS

REVERTING to fact from fiction, it is desirable to bring into its historical place the overseas activity of Portugal, which dates from the beginning of the fifteenth century— that century to which, in its final decade, belong two of the most important incidents in our history on the side of exploration—the crossing of the western ocean by Christopher Columbus, and the completion of the sea-route to India around the Cape of Good Hope by Vasco da Gama.

Portugal had had to fight long for the firm establishment of its independence. Toward the end of the fourteenth century Castile was still a powerful enemy within the peninsula; the Moors were still the objective of crusading fervour without. In and after 1385, however, the Portuguese, aided by English arms, overcame the Castilians, and by the end of the century an honourable peace with Castile and a close union with England by treaty were achieved. In 1415 a Portuguese fleet, again with English armed support, captured from the Moors the port of Ceuta, which thus became the first African outpost of Portugal.

The political power of Portugal established, its geographical position gives obvious reason for the direction in which the country took the lead in overseas development. Portugal was the European territory (of commercial consequence) furthest removed from the overland routes to the East, and landward trade even into Europe must follow difficult routes, even if political conditions were favourable. The Atlantic seaboard, on the other hand, offers good harbours, and actually trade with middle Europe and England followed the sea. Moreover, the crusading spirit was present to be directed into fresh channels of exploration.

Its first and principal director was Prince Henry, surnamed

the Navigator (1394–1460), fifth son of King João (John) I of Portugal by Philippa, daughter of John of Gaunt. It is tempting to speculate upon the psychological implications of the alliance between Portugal and England. Prince Henry has a very great name in the history of geography: he was truly a 'maker'. He was not an explorer himself, but he inspired and instructed the sea-captains who voyaged from Portugal during his lifetime, and the inspiration, at any rate, survived him. He began his geographical work about 1415. In 1419 he became governor of the southern province of Algarve, and at Sagres, in that province, he lived for most of his life. He was a patron of learning, not only geographical; so far as concerns geography and navigation, he provided instruction in cartography and astronomy, including the use of the appropriate instruments, and he appointed his teachers with a fine disregard of political sentiment, for some of his mathematicians were Arabs. There seems every reason to believe that he had in mind the possibility of reaching the Indies by sea not only around Africa but also across the western ocean. The extension of geographical knowledge was not, of course, his sole object. He aimed at, and achieved a large measure of, the expansion of Portuguese commerce. His Christian mind desired to afford spiritual help to the natives of the lands visited by his people, and incidentally he hoped for the establishment of communication with Prester John. After the Portuguese voyagers had for some years carried on an active slave trade from West Africa, Prince Henry prohibited it.

The principal incidents of Portuguese voyaging during and after Prince Henry's life may be tabulated thus, the captains' names being bracketed:

1415.   Grand Canary (João da Trasto).
1418–20. Porto Santo and Madeira (João Gonçalvez Zarco).
1427–31. The Azores (Diogo de Seville and Gonçalo Cabral).

The above islands were previously known, at least in part,

but there was no systematic colonization or trading with them till after this time.

1434.   Cape Bojador passed (after many attempts) (Gil Eannes).
1441–2. Cape Blanco and Arguim (Nuno Tristam).
1445.   The Senegal River and Cape Verd (the same and Diniz Diaz).
1446.   Sierra Leone (near) (Alvaro Fernandez).
1456.   Cape Verd Islands (Alvise Cadamosto).

Cadamosto and Diogo Gomez, about this time, penetrated the Senegal and Gambia rivers for considerable distances.

1482.   The Congo River (Diogo Cão).
1485–6. Cape Cross (Damaraland) (the same).
1487–8. The Cape of Good Hope discovered and rounded: Great Fish River (south-east coast of present Cape Province).
1497–8. The Cape route to India (Vasco da Gama), which falls for consideration later.

It will be seen that the longer voyages to and beyond South Africa took place later than the death of Prince Henry (1460). But his, to all appearance, was the groundwork of this super-structure.

Cartography, under the impulse of discoveries on the one hand and the republication of Ptolemy's work on the other, reached in the fifteenth century a stage transitional between the primitive and the fully developed. The earlier extensions of the *portolano* maps to cover the world were not con-spicuously successful outside the Mediterranean area. On the other hand, some of these maps, very naturally, recorded the Portuguese discoveries more promptly than the maps which were based upon those of Ptolemy. Of his work there were seven editions in this century. About 1470 a German monk known as Nicolaus Germanus produced a finely illu-minated edition hand-drawn, with additional maps. Fran-cesco Berlinghieri's edition issued at Florence in 1480 contained printed maps, including new maps of France, Spain, Italy, and the Holy Land. A conical projection appears in use about 1480. The first conception of the modern atlas now comes into being.

Ptolemy's error concerning the shapes of India and Ceylon died hard, and the Portuguese discoveries did not soon dispose of his misconception of the form and extent of Africa. The necessity of showing the known world persisted

as a convention: the map of Fra Mauro already mentioned (1457) did this to the notable detriment of the shape of Africa, though the cartographer well used the information of overland travellers to the east, down to Nicolo de' Conti, a Venetian merchant of his own time, who, travelling in India and as far east as Java, was away from his own city for twenty-five years from 1419. There was also some measure of com-

FIG. 12. Outline of Fra Mauro's World. (In the original the south is at the top.)

munication between Italy and Abyssinia at this period, and Fra Mauro used information acquired from natives of that part of Africa. The circumference of the earth was generally underestimated. The eastward extension of Asia to 180° or thereabouts, was common belief, though a Genoese map of 1457, if its proportions are to be relied upon (for there is no graduation) would have brought the coast of China only to 136° E. by our reckoning. Maps produced in Europe quite early in the century showed knowledge of the North Atlantic Ocean as far as the coast of Greenland. On the other hand, islands associated with various myths back to that most ancient story of Atlantis appeared commonly in the ocean farther south.

Chapter XI

THE GERMAN SCHOOL (1500–50)

IN 1487 Bartholomeu Diaz, as we have seen, rounded the Cape of Good Hope and reached the Great Fish River. After ten years of preparation the King of Portugal sent an expedition to India by this route, which, under the leadership of Vasco da Gama sailed from Lisbon in July 1497. On November 22 he reached the Cape, rounded it, sailed along the east coast of Africa as far as Melinde, and thence, with the aid of an Arab pilot, reached Calicut on May 20, 1498.

In the meanwhile, Isabella of Spain was finally convinced of the possibility of reaching the Indies by a route shorter than that round the Cape, and on August 3, 1492, with her patronage, the Genoese Christopher Columbus set sail from Palos westward. He passed by the Canaries and after a long journey of mingled hope and despair ultimately sighted land on October 11, deemed by the explorer even till his death to be a part of south-eastern Asia. In his first voyage Columbus landed on one of the Bahama Islands and discovered Hispaniola and Cuba. In his second voyage (1493–6) he discovered Jamaica and explored the south coast of Cuba; in his third (1498–1500) he found Trinidad and observed fresh water out at sea, and thereby assumed the existence of a continent, a sound conclusion, for in fact he had discovered the mouth of the Orinoco. In his last voyage (1502–4) he intended to sail through the archipelago, and so circumnavigate the world, but he reached Cuba and for four months explored the coasts of Honduras (which he believed to be part of the mainland of Asia) as far as Darien. Then he returned to Spain and died shortly afterwards.

The true outlines of the New World were quickly revealed during the early sixteenth century. John Cabot, sailing from Bristol with ships provided by Henry VII, traced the coastline from the south of Greenland to Virginia, and a few years later (1500–1) Gaspar Cortereal sailed along the east coast of Greenland and visited Newfoundland. Francis I of France sent out Jacques Cartier on four voyages (1533–43) and he sailed up the St. Lawrence River and made an abortive attempt to establish a colony on the site of Montreal. In South America, Pinzon, in command of an expedition in 1499–1500, crossed the Equator and discovered the Brazilian coast at Cape San Agostinho, explored three hundred miles of coast-line, reached the mouth of the Amazon, and possibly even passed the mouth of the Rio de la Plata. Some three months later Cabral, on a return voyage from India, was drifted westward across the South Atlantic and also struck the coast of Brazil. The Italian Amerigo Vespucci was immediately commissioned by Portugal to explore this coast thoroughly, and he reached as far south as lat. 34° (1501) and discovered the island of Georgia which he took to be a part of a southern continent.

In 1513 Balboa saw the Pacific Ocean. Quickly followed the exploration of Mexico and South America by the Spanish conquistadores. Cortes reached the Gulf of California in 1533, and by 1571 it was possible for Ortelius to produce his excellent map of Mexico. Pizarro rapidly conquered the Inca Empire, and in 1535 added Chile to the Spanish possessions. Thus the whole coastline of South America north of Santiago, and North America south of San Francisco, was explored before the middle of the sixteenth century. In addition to coastal exploration and the Spanish conquests in South America, Francisco de Orellana in 1541 journeyed right down the Amazon from Quito; in 1539 Hernandez de Soto, lured by the prospect of treasure and precious metals, undertook an unsuccessful expedition into Arkansas; and finally abortive attempts were made to establish French Huguenot colonies between 1555 and 1564 in Brazil and Florida.

Problems still remained, however, regarding the extremities

of the New World. The configuration of the Arctic shores of North America was gradually worked out by the relentless search for a north-west passage to India, while the southern extremity was rounded by Magellan in 1519, who believed Tierra del Fuego to be a part of a southern continent, an idea which was later disproved by Drake in his, the second, circumnavigation of the world (1577–80).

In the Far East trading voyages and exploration were undertaken almost exclusively in the sixteenth century by Spanish and Portuguese. Portugal dispatched voyagers to Sumatra, Java, and the Philippines in the early years of the century, in 1512 to the Moluccas, in 1516 to Canton; and in 1530 a Portuguese embassy was established at Peking.

Geographical conceptions on the eve of the age of discoveries in the sixteenth century had made little advance on the philosophical speculations of the Greeks. The flat-disk theory of the shape of the earth was generally held by the religious thinkers of the Middle Ages (e.g. Fra Mauro, 1459), though we have already met with a more enlightened few who believed in its sphericity, such as Bacon. To these should be added Cardinal d'Ailly in his *Imago Mundi* (1415), while as far back as the middle of the thirteenth century John Holywood (Sacrobosco) had maintained the rotundity of the earth in his *De Sphaera Mundi*. That this view was generally accepted in the fifteenth century is evidenced by the making of globes of which examples have been given. But the Ptolemaic tradition permeated the cartography of the Renaissance, which was not entirely freed from it until the eighteenth century. D'Anville, who finally purged cartography of this tradition, referred to the measurement as 'the greatest of geographical errors and the seed of the discovery of the greatest geographical truths'.

Medieval ideas of the distribution of land and water were largely speculative. Aristotle taught that around the earth were four concentric envelopes—earth, sea, air, and fire, which were not everywhere regularly superposed, for in places the land lay adjacent to the air, and elsewhere the sea filled cavities in the land. To account for the distribution of the known lands, the centre of the spherical earth was slightly displaced from that of its aqueous envelope so as just to protrude through it—though the protrusion was almost tangential to the surface of the aqueous envelope, thereby avoiding instability. This early medieval idea is probably associated with the form of the land on maps of the time (*mappemonde*), for according to the hypothesis there would be one round land mass such as is shown on medieval maps. But many early philosophers speculated on the existence of other unknown lands, and a more common conception was that on the spherical earth, the sea lay in cavities, separated by land masses, though the distribution of the latter was long disputed, until their actual discovery and the advance of science. Copernicus combated the idea of the two spheres of water and earth, for the preponderance of the former would make the globe unstable, the seas could not possibly approach so close to Egypt because it was situated in the middle of the (known) lands, and further the existence of islands could not be accounted for on this hypothesis because, according to it, the sea should become deeper outwards from the borders of the continent. On these grounds Copernicus concludes that there is an earthen core, with a water-covering concentrated in the hollows of its surface.

The old world we have found to be given in the medieval wheel maps a circular or oval shape, completely surrounded by water; the land is usually divided into three continents by the Mediterranean, Red Sea and Nile. Early science postulated the existence of other habitable lands or *oekumene* in the southern hemisphere, or in the same latitude as the known world in the northern hemisphere. But before the discoveries there was no knowledge of the actual existence or extent of these lands; theory outran facts. From these speculations

arose problems which were soon solved with the new discoveries. Thus the Torrid Zone was believed by some to be
entirely water and by others to be too hot for habitation—
though Ptolemy extended Africa across the Equator. Moreover, if the earth were a sphere, how could people living in
antipodal lands walk upside down like flies on a ceiling?

Medieval cartography (except for *portolani* charts) was un-
scientific. No attempt was made to use projections or to use
data of longitude and latitude to plot the exact position of
places. The discoveries proved that the world was more
extensive than had formerly been believed and new projections were necessary. Records to fix the position of places
were also essential. Thus the first phase in the development
of cartography in the Renaissance period is the correction and
augmentation of the Ptolemaic records, and the invention of
new projections.

In the first half of the sixteenth century Germany was the
principal centre of development of both mathematical and
descriptive geography. The school of cartographers was
fathered by Peurbach and his pupil Regiomontanus at the end
of the fifteenth century. They collected and corrected the
astronomical observations of Ptolemy and Regiomontanus,
and published ephemerides for the period 1474–1506 which
were used by Columbus. Tables of latitude and longitude,
maps, and globes were prepared by most of the cartographers,
but outstanding in their contribution to the study of map
projections are Jean Werner and Peter Apian. Martin Behaim, a pupil of Regiomontanus, made a globe at Nuremberg
in 1492, which gives the best impression of the conception of
the distribution of land and water on the eve of the discoveries. Obsessed by Ptolemaic tradition, he showed the
same south-eastern extension of Africa, with Madagascar and
Zanzibar as two large islands. He retained the Ptolemaic idea
of the non-peninsular form of India and the gross exaggeration
of Ceylon and the Malay Peninsula. The trend of the west

coast of Africa, the Gulf of Guinea, and the Cape of Good
Hope are shown. The east coast of Asia almost coincides with
the true west coast of North America, the East Indies sprawl
over the South Pacific and Cipangu (Japan) lies immediately to the west of central America. In the midst of the

FIG. 13. The World according to Martin Behaim. (Actual land outlines
shown unshaded.)

Atlantic are shown the two mythical islands of Antillia and
Brandon.

To Martin Waldseemüller, professor of cosmography at
St.-Dié, is credited the christening of the New World,
America. The name is given to the southern land-mass which
is shown on his globe (1507) as an island, and it is separated
by a channel from a similar land-mass to the north. In
Ruysch's map (1508) South America appears as a separate
land-mass, the coast of North America is identical with that
of north-east Asia, and the West Indies appear as an archipelago. Of Schoner's globes, that published in 1520 shows
three islands in the New World, Litus Incognitum in North
America, a central archipelago, and a southern island,
America. But in his last globe (1523) North America is
separated from Asia by a long narrow strait, a definite north-

west passage, South America has its correct shape, and the central archipelago has disappeared.

As indicative of the knowledge of the distribution of land and water, Sebastian Munster's map (1532) may here be noted. He shows no southern continent. Africa extends to lat. 35° S., India is still small, Malaya much exaggerated, and also Taprobane (Ceylon). The Chinese coast is in longitude 190° to 120° E. of the standard meridian which passes through the Madeira islands. He shows three channels penetrating the New World; one to the north of North America, another through Central America, and a third to the south of South America, with Tierra del Fuego forming part of a southern land-mass.

Long ocean voyages and the exploration of new lands resulted in the improvement in methods of navigation, the careful observation of winds, weather, and currents, and the accumulation of vast stores of information relating to the commerce and peoples of foreign lands. At first all these data were compiled by recorders, though systematization and correlation on a scientific basis came very slowly. The three greatest compilers of records of voyages of exploration in the sixteenth century are Hakluyt, the Italian Ramusio, and De Bry. Richard Hakluyt arranged the translation and publication of Mendoza's *China* (1589), Pigafetta's account of the Congo (1597), and a work on Africa by a Moor, Leo Africanus (1600). He collected records of voyages from all the great explorers, which were published in his *Divers Voyages touching the Discovery of America and the islands adjacent to the same* (1582) and *The Principal Navigations, Voyages and Discoveries of the English Nation made by Sea or over-Land to the most remote and farthest distant quarters of the Earth* (1588). After his death, Samuel Purchas used his manuscripts for the preparation of his *Hakluytus Posthumus or Purchas his Pilgrimes*. Hakluyt was a mine of information and was frequently consulted by intending explorers or traders.

He supported Raleigh's Virginia colonization enterprise, and was a director of the Virginia Company, and later became adviser to the East India Company. He never published all his work in one treatise, but compiled masses of facts which he used in advisory capacity. He was in short 'a geographical society in his own person'.[1]

The two chief representatives of theoretical geography of the period are Peter Apian and Sebastian Munster, and for a hundred years theirs were the standard works, the former for his popular exposition of astronomy and mathematical geography and the latter for his brilliant descriptive geography, modelled on Strabo. Pierre Bienewitz, or Peter Apian, born in Saxony in 1495, was an astronomer and cartographer. In addition to making maps and globes, he published two works, *Astronomicon Caesarem*, an astronomical treatise, and *Cosmographicus Liber*. He, like Schoner in his *Luculentissima Descriptio*, modelled his work on the *Cosmographiae Introductio* of Waldseemüller, and in fact much of his material is taken from Schoner's work; yet it became popular and the latter was forgotten, mainly because by means of charts and diagrams Apian popularized an otherwise difficult subject. The *Cosmographicus Liber*, first published in 1524, was subsequently edited and enlarged by Gemma Frisius under the title of *Cosmographia*. The book in its original form dealt almost exclusively with those aspects of geometry and astronomy which are essential to geography. Long lists of latitudes and longitudes are given for numerous places, and what descriptive geography it contains was subsequently appended by Frisius.

The Ptolemaic distinction between geography and chorography is clearly stated. Chorography 'describes and considers places separately—without consideration of comparison between themselves, or with the world as a whole'.

[1] G. B. Parks, *Richard Hakluyt and the English Voyages* (American Geographical Society, Research Series, 1928).

The aim of chorography, he writes, 'is to paint and describe a particular place, as a painter would paint an eye or an ear or other parts of a human being'. Geography, on the other hand, is the general description of the whole earth, and bears the same relation to chorography as the painting of a human being to an eye or an ear.

The earth is shown at the centre of the universe, the sun and planets revolving round it. The earth is divided into five zones, torrid, temperate (between the tropics and polar circles), and frigid. Climata are defined as the spaces between parallels of latitude at intervals of a difference of half an hour in the length of their longest day, and each climate is named after a principal feature in it, a town, river, or range of mountains. The lands are of four forms, islands, peninsulas, isthmuses, and continents, with simple diagrams of each. Then follow diagrams of hands and feet to serve as a basis for linear measurement. Very short notes are given on each continent, and finally there is a long list of towns for each country, with their latitudes and longitudes taken from Schoner and Ptolemy. It is to this section that Frisius has appended his descriptive notes. Sebastian Munster (1489–1552) is the best representative of the school of German geographers, and his great work *Cosmographia* was the standard reference for over a hundred years. Born at Ingelheim (between Mayence and Bingen) in 1489, he studied at Heidelberg and Vienna, and later taught at the former university (1524–7). In 1529 he was appointed to a chair of Hebrew at Basle and there remained until his death in 1552. Munster made important contributions to cartography. He prepared an edition of Ptolemy published at Basle in 1540, four re-editions of which appeared during the next twelve years. He attempted to improve cartographic methods, by using, for the first time, a small compass, the forerunner of the prismatic compass, for a simple triangulation survey of a small area round Heidelberg. His method was published and he suggested a plan for an accurate survey of the whole of Germany. A few points were to be fixed astronomically and then detail was to be filled in with the compass. Different parts of Germany were allocated to various authorities with a view to preparing a reliable map. The scheme, however, did not mature. Munster was also interested in the discoveries, and he produced a world map in 1532 the features of which have already been noted.

But *Cosmographia* was his great work. He gathered material from many contemporary authorities, so that it is a compilation rather than a carefully arranged treatise, and history and genealogical tables take up a large part of the book. Mathematical and physical geography are almost excluded in his treatment. That the earth is round is taken for granted. He declares that the earth's crust suffers changes through floods and the work of rivers. Many lands have been flooded since the Deluge, and mountains and valleys formed by rivers where the land was formerly flat, these observations being based on his knowledge of floods in Holland. He also mentions earthquakes, the 'central fire', the character of rocks, the nature and distribution of metals and methods of mining.

The bulk of the book deals with human and political geography on a regional basis, and a few quotations will illustrate the general character of the descriptions. In the section on the British Isles, the Tweed and Solway are mentioned as separating England and Scotland.

'There are also two other important rivers, one called the Humber, the other the Thames, on the banks of which is situated the royal city of London formerly called Trinovantum. In our day it is a great commercial city with much trade, as large boats can reach it. . . . One and a half leagues from London to the east is Greenwich where usually dwell the English kings. From here boats ascend as far as London; these are not drawn by horses, but are driven by the wind, or sail on the tide which rises and falls twice each day.'

An interesting comparison is made between Spain and France.

Gaul is fertile owing to its plentiful rivers, and Spain uses irriga-tion by taking water from the large rivers in ditches. It is not affected by cold north winds as Gaul. . . . The country of Spain is larger but not so thickly peopled. It is richer in gold but not in merchandise, and not so much revenue is collected as in Gaul. There is scarcely any soil in Gaul which is not useful, but in Spain there is much desert and uncultivated land.'

Russia is described as 'very flat with no mountains though forests and marshes occur everywhere, with fine rivers. The principal city of the country is Moscow, which is twice as large as Prague in Bohemia. The fortifications are made of wood as in other towns of this country.' Writing of Transyl-vania in his section on eastern Europe, he says, 'This region is surrounded on all sides by great and lofty mountains, as a town is surrounded and fortified with boulevards and walls'.

Most of the book, however, is devoted to a detailed account of Germany, the limits of which he takes as being the Rhine and Danube, though since classical times he notes that the German language has spread beyond these limits. 'The Black Forest', he writes, 'is filled with pines and has many great tall trees; there are also high mountains. . . . It is inhabited in all parts except for several high desert mountain summits.' The timber trade is said to be very important and wood is floated down the Danube to Ulm. He says of Franconia that it 'is enclosed by dense, dark, almost impenetrable forests and gloomy mountains. Inside, the land is flat and uniform, adorned with many isolated towns, castles, fortresses, and villages. The forest called Hercynia and the mountains surround it as with a natural wall. The Maine (a navigable river), Sal, Tuber, and Neccar pass through the region and the valleys are broad and deep, from which are obtained excellent wines. These are widely distributed because of their high quality.' The fifth and last part of the book deals with Asia, Africa, and the new lands. All the material here is second-hand, and the descriptions are much inferior to the section on Europe.

Munster's work is almost entirely descriptive with much irrelevant detail, while the mathematical and physical or general geography is omitted. Apian's *Cosmographicus Liber* formed a useful companion volume to it. But Munster was a keen observer and a good writer and his work became the standard. 'So completely did the volume resulting from the insight, learning and energy of Munster meet the demand of the time, and so thoroughly did it establish itself, that in enlarged form it remained in use until after 1650, going through forty-six editions and appearing in six languages.' [1]

[1] A. H. Gilbert, 'Pierre Davity: His Geography and its Use by Milton', *American Geographical Review*, vol. vii, 1919.

Chapter XII

THE FLEMISH SCHOOL (1550–1650)

DURING the next hundred years, the chief centres of cartographic activity were in the Netherlands, while in 1650 there was published at Amsterdam the excellent geography of Bernard Varenius, which summed up the knowledge of the earth at that time and was a standard work until the end of the eighteenth century.

The outstanding features in the development of cartography in the sixteenth century are, first, the invention of projections suited to the mapping of the whole world; secondly, the collection and correction of astronomical records; thirdly, the accurate survey of relatively small areas; and fourthly, the co-ordination of all maps for the production of reliable maps of the world.

The Flemish school and its contemporaries effected enormous advances in cartography and did much to free it from Ptolemaic errors. Its two chief representatives are Gerhard Kremer, or Mercator (1512–94), and Abraham Ortelius (1527–98). Mercator, born at Rupelmonde in Flanders, studied at Louvain where he met Gemma Frisius, and later he was a pupil of Apian's at Ingolstadt. With the help of Frisius he founded a geographical establishment at Louvain for the manufacture of astronomical instruments, the collection of records, and the preparation of maps. During the period 1537–40 he surveyed and produced a map of Flanders, the first map of a large area to be based on a detailed survey. This was followed by other large-scale surveys, using the new methods of triangulation which were now coming into general use (see Chapter XIII). Maps were prepared of most countries in western Europe in the second half of the sixteenth century. Of these maps the most noteworthy are: Gastoldo's map of Spain (1544), Philip Apian's map of

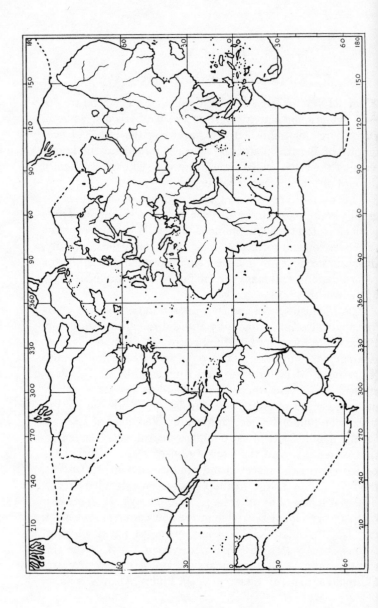

Bavaria on a scale of 1/144,000 (1568), and La Guillotière's map of France published in 1613. In 1538 Mercator produced his first world map, which shows marked Ptolemaic influence. In 1552 he was appointed to a chair of cosmography at Duisburg, and he published in 1554 a map of Europe in six sheets, in which he begins to emancipate himself from Ptolemy, for the length of the Mediterranean, given as 62° by Ptolemy, is reduced to 53° in this map. In 1563 he published a map of Lorraine, but his greatest work was his planisphere of the world for navigation (1568). On this map there are two very important new features. First, the map is drawn on a new rectangular projection, which preserves correct shape for small areas and gives true directions. Ptolemy's modified conical projection was suited to a map of Europe, but not for the world, as it greatly distorted the southern hemisphere. Moreover, the globular projections of the German school resulted in great distortion and did not give correct direction. This latter defect, to the advantage of the navigator, was remedied by what is now known as the Mercator projection. Secondly, Mercator adopted a new prime meridian. He had in his earlier maps adhered to Ptolemy's selection of the Fortunate Islands, the position of which were but vaguely known. The magnetic meridian, discovered by Columbus, was considered to be an excellent natural line suitable for the origin of longitudes. In 1553 Mercator decided to adopt it, and its position he took to be marked by the Azores on the evidence of navigators' records. Yet, owing to the lack of accurate records in the Far East and the New World, and reliance upon Ptolemaic records for the former, these two land masses were still much exaggerated in size—the greatest errors being on the west coast of the Americas and the east coast of Asia and Africa. Further, beyond explored areas, Mercator let his fancy run riot. Africa he filled with names from Ptolemy, and added matter from the even more imaginative map of Africa by F. Pigafetta

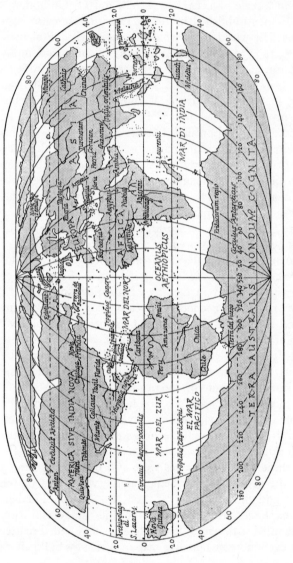

FIG. 15. Ortelius's World-map, 1570.

(1591). In unknown Asia he jumbled Ptolemaic place-names and features borrowed from Fra Mauro's map, which Mauro in turn had taken from Marco Polo's travels. Lastly, around the south pole Mercator shows a vast southern continent which includes Tierra del Fuego, extends north towards South Africa, and partly overlies the site of Australia.

A few years before his death Mercator began to prepare an atlas with detailed descriptions of each map. He died, however, in 1594, and the work was completed by his son Rumbold in 1595. A more famous atlas, the *Theatre of the World*, was produced some years previously (1570) by Abraham Ortelius, a compilation which included fifty-three maps in the first edition, and over a hundred in 1595, and for the preparation of which Ortelius had consulted eighty-seven authors. He launched into this great work through the inspiration of Mercator, with whom he travelled in 1560.

Jodocus Hondius (1563–1611), brother-in-law of Rumbold Mercator, continued the family tradition; he is chiefly known for his map of the world (1595), which shows for the first time the course of Drake's voyage round the world (1577–80). In this map, as in Mercator's, continents are shown in the South Seas and in the Arctic.

The extent of the knowledge of the distribution of land and water at the end of the sixteenth century is shown on Mercator's world map (1569), and better still on Wright's world map of 1600, which is the earliest English example of a map on the Mercator projection. Conjectural geography is rigidly excluded from the latter and only explored coast-lines are shown. There are two areas of unexplored territory —the north of the Pacific and the connexion between America and Asia, and the site of the continent of Australia. These were to be the centres of the chief discoveries during the next two hundred years. The outlines of the old and new worlds are fairly well known, though, through the lack of records,

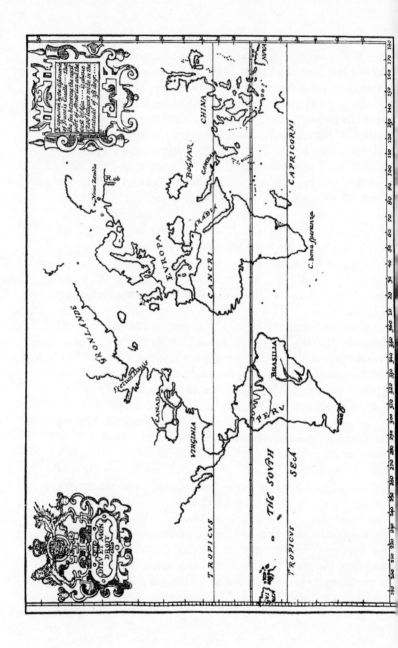

It appeared by the discoverie
of Francis Gualle, That
America on the northwest
coast doth not... is about
1200 league wide in the
latitude of 38 deg.

DEV ET MON
DROIT

GRONLANDE

Fretum Davis

Nova Zembla
R. ob

EVROPA

BOGHAR

CHINA

CANADA

VIRGINIA

Candisa

ARABIA

CANCRI

TROPICVS

PERV

BRASILIA

THE SOVTH

SEA

C. bona speranza

CAPRICORNI

TROPICVS

NOVA

their shapes are often distorted, mainly in an east-west direction, owing to Ptolemy's faulty records of the Old World and the difficulties of taking accurate measurements of longitude.

After 1600 England began to take an active interest in the East Indian trade, for in that year the East India Company was founded. In 1602 the Dutch East India Company was also formed. James Lancaster was dispatched to the East in 1600 on behalf of the English company, and he laid the foundations of English commerce in the Spice Islands, visiting Java, Sumatra, and the Nicobars. It was subsequent to the defeat of the Portuguese off the coast of Java by a Dutch fleet under Cornelis Houtman that the Dutch company was inaugurated. So ended the Portuguese and Spanish dominion in the Malay Archipelago.

Knowledge of central Asia in the sixteenth and seventeenth centuries was based upon the vague reports of medieval and contemporary travellers, missionaries, and embassies. Between 1558 and 1579 traders in London established trade connexions with the lands round the Caspian Sea and embassies visited Bokhara, Persia, and Russia. In 1579 Anthony Jenkinson went to Persia as the representative of Elizabeth and gathered much valuable information; and he was later followed by Sir Anthony and Robert Shirley. Many travellers also visited Asia Minor, Syria, and Persia in the seventeenth century, but added little to the store of true geographical knowledge.

Missionaries were the only sources of information of China and Tibet down to the end of the seventeenth century. Tibet, visited in 1325 by Friar Odoric, was not entered again by a European till 1624. P. Tachard (1685-7) journeyed in Cochin-China and Tong-King and collected many valuable astronomical observations, which helped to prove the great longitudinal errors of Ptolemy. During the seventeenth and eighteenth centuries the Jesuits, permitted to enter China in

1553, collected many data. Meanwhile the Russians were penetrating eastwards along the north coast of Asia. Hunters reached the Sea of Okhotsk in the early seventeenth century and a few years later the Amur was navigated to the sea. In 1768 an organized exploration of the whole of the Russian empire was undertaken.

The two great problems of the distribution of land and water to which attention is mainly directed during this period are the north-east and north-west passages, and the distribution of land in the southern hemisphere. Most outstanding features of oceanic discovery are the voyages of Tasman and Cook, the latter heralding the period of scientific exploration; and the search for a northerly passage to the East Indies.

In the early stages the search for a passage around the north of America and Asia was mainly commercial, but later voyages were undertaken solely for the sake of discovery and additional knowledge. The Portuguese possibly followed in the wake of Cabot and Cortereal, but of their early voyages there are no existing records. But serious exploration was undertaken by English explorers at the end of the sixteenth century. Martin Frobisher (1576-8) reached eastern Greenland and the southern coast of Greenland; John Davis made three voyages (1585-7) in the course of which he explored the Strait named after him and sailed along the west coast of Greenland as far north as 72° 41'. Henry Hudson in 1607 reached the east coast of Greenland in latitude 73° N., investigated ice conditions between that country and Spitsbergen, and discovered Jan Mayen island. After a second voyage to search for the north-east passage, he turned west again and discovered the Hudson river, strait, and bay. In 1610 he was set adrift by a mutinous crew in Hudson Bay and was lost. Thomas Button in 1612 reached the west coast of Hudson Bay and expressed the opinion, which was long held, that the north-west passage opened from this coast. William Baffin (1615-16) penetrated three hundred miles beyond Davis's northern limit and took valuable magnetic observations, in the course of which he discovered the greatest known variation of the compass in Smith's Sound. It was not till 1770 that the passage to the north of Hudson Bay was discovered by Samuel Hearne, and in 1789 Alexander Mackenzie reached the river which bears his name. In 1776 Cook approached Bering Strait from the Pacific and was stopped by ice at 70° 41′ N.

The first English expeditions to search for a north-east passage were organized by Sebastian Cabot and led by Sir Hugh Willoughby and Richard Chancellor. Chancellor reached the White Sea and penetrated to Moscow, and the Muscovy Company was established as a result. Shortly afterwards (1556) this company sent Stephen Borough to explore in north Russia, and thus information was obtained of the Kola region and Novaya Zemlya. Dutch expeditions were led by Olivier Brunel (1582) and William Barents (1594 and 1596), the latter discovering Spitsbergen and Novaya Zemlya. In the next two centuries the Russians explored most of the north coast of Siberia. In 1735 Chelyuskin rounded the promontory named after him and in 1728 and 1740 Bering explored the Bering Strait. The development of whaling in the Arctic seas stimulated polar explorers, among whom the most prominent is William Scoresby, a captain of a whaling ship, who penetrated north to 81° 12′ 42″ N. in 1806, and explored the east coast of Greenland from 75° to 69° N. in 1822.

The concept of a southern continent in temperate latitudes, south of the Torrid Zone, was originally based on Greek speculations, and later possibly on vague rumours reaching the West from the Far East from travellers such as Marco Polo, while French and Portuguese ships possibly stumbled across the northern shores of Australia in the early sixteenth century. The existence of an oecumen to the south of the Equator, separated from the known lands by an ocean in the

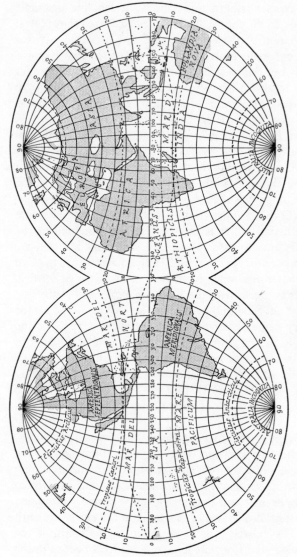

FIG. 17. De Witt's World-map, 1700.

Plate IV. (a) Astrolabe (1537) by Georg Hartmann of Nuremberg

(b) Azimuthal quadrant of Tycho Brahe,
c. 1587, for measuring altitudes

Torrid Zone, was postulated in classical times as we have seen, and later by Pomponius Mela, while Ptolemy showed the Indian Ocean as a *mare clausum*, with a vast continent engirdling the south pole. The idea is perpetuated, mainly on the authority of these sources and the travels of Marco Polo, in the world maps of the sixteenth century, of which Mercator's may be taken as typical. His southern continent, called Magellican, includes Tierra del Fuego, and stretches continuously to the site of Western Australia—as yet undiscovered. The existence of this vast imaginary continent was an obsession to most cartographers and explorers in the sixteenth and seventeenth centuries, and every new discovery in the southern hemisphere was deemed at first to form a part of it. On some early sixteenth-century maps South America is shown as a northerly extension of it; after Magellan's voyages, it was pushed south to Tierra del Fuego, until Drake proved Callao to be an island. De Quiros set sail from Callao in 1605 for the south land and reached the New Hebrides, which he declared to be a part of it. The conception of Java as part of the continental land died hard, even after it was circumnavigated at the end of the sixteenth century, and in Homan's map of 1716 even New Guinea still figures as a peninsula. Dutch voyagers in the south seas of the Pacific perpetuated the same idea.

These erroneous conceptions were ultimately dispelled by the voyages of Tasman, Dampier, and James Cook. The exploration of Australia by Europeans begins with the opening years of the seventeenth century, and by 1665 the Dutch had explored and charted most of the western and northern seaboards. In 1642 Abel Janszoon Tasman sailed from Batavia to Mauritius and thence south-eastward. He reached the south and east coasts of Tasmania, which he called Van Diemen's land, and passed along the western coast of New Zealand, which thereby assumed the position formerly held in turn by New Guinea and Australia as a northern peninsula of the southern continent. William Dampier (1697–1701), in his voyages across the Pacific, around the known shores of Australia, and from the Cape of Good Hope across the south of the Indian Ocean, did much to dispel this idea, in addition to gathering important data and preparing excellent records of wind directions. He was the first definitely to prove New Guinea to be an island, for Torres's voyage of 1606 was unknown till late in the eighteenth century, and Tasman had returned by its north coast. Yet even in 1721 a certain Jacob Roggeveen, leading an expedition for the Dutch East India Company, regarded land which he found to the south of Tierra del Fuego, and the island of Samoa, as promontories of the mythical continent. Finally James Cook laid to rest this age-old myth and definitely proved Australia and New Zealand to be islands. He also proved again that New Guinea is not a part of Australia. On his second voyage to the south seas (1772) he sailed from the Cape of Good Hope to New Zealand, passing within the Antarctic Circle twice on the way, and thence sailed three times across the Pacific, in the course of which he discovered many of the island-groups of the south seas. Finally he sailed from New Zealand to Tierra del Fuego and the Cape of Good Hope.

The world-wide ocean voyages and the records of navigators, the need for careful observation of winds and currents, the reports and observations of missionaries and travellers, all resulted in an enormous accumulation of facts concerning the earth's surface. Observations of currents were made easily by the Portuguese. They were familiar with the warm Guinea and Gulf Stream currents in the early sixteenth century, Cabot had observed the cold Labrador current in 1497, and the Humboldt current was known three hundred years before Humboldt wrote about it. Columbus in his third voyage noted that the waters of the sea move from east to west, 'like the sky', the first record of the common move-

ment of wind and water near the Equator. In 1578 it was
noted that it was much colder in latitude 62° in North
America than in Norway 8° farther north. Captain James,
who was in Hudson Bay during the winters of 1631 and 1632
in 52° N. noted the strong contrast between its climate and
that of London. On one of Barents' expeditions Novaya
Zemlya (lat. 76° N.) was found to be much colder than Spitsbergen (80° N.). As regards winds, the observation of the
general wind systems was an easy matter on the open ocean
as compared with the calms and local winds of the Mediterranean and the strong variable winds of the North Sea lands.
The simplicity of the wind systems was early appreciated
and their direction taken advantage of for ocean voyages.
Thus the North-east Trades were used for the outward
journey to the New World by the Spaniards, and the Westerlies for the return. They sought and found, after twenty
years' effort, similar winds in the Pacific to assist the long
journeys from the Philippines to Mexico, by returning from
Manila with the help of the Westerlies in latitude 35° N. The
Monsoons, regularly used by the Arabs for sailing between
India and east Africa, were also soon familiar to the Portuguese and their successors.

Observations were also made of the different peoples;
e.g. Frobisher in his first expedition made contact with the
Eskimos of whom he writes, as quoted by Hakluyt: 'They be
like Tartars with long black hair, broad faces and flat noses
and tawney in colour.' During this period astronomy made
great strides thanks to the works of Copernicus (1473–1543),
who explained the alternation of day and night by the rotation
of the earth on its own axis, and its movement round the sun,
as opposed to the geocentric system of Ptolemy; while Kepler
(1571–1630) later defined his laws relating to the movement
of the members of the solar system.

As yet no authoritative geography had been published
which satisfactorily combine (1) general, mathematical, and

physical geography in the light of these recent researches and records and (2) the more detailed exposition (chorography) of the geography of countries on the basis of the vast accumulation of facts appertaining to newly discovered lands; for Munster's work dealt almost exclusively with Europe. It is almost certain that had Bernard Varenius lived he would have completed such a work. The only volume he published on general geography is almost modern in its outlook and the vision of the author is only hindered by the lack of truly scientific data.

Bernard Varenius was born in 1622 at Hitzacker, a small town on the Elbe near Hamburg. In 1640 he entered the gymnasium at Hamburg and studied philosophy, mathematics, and physics. After three years he went to the university at Königsberg to study medicine and, after staying there a year and a half, dissatisfied with the teaching, he went to Leiden to pursue the same work. In 1647 he took a post as a private tutor to a family in Amsterdam. After failing to obtain an appointment at the gymnasium at Amsterdam, he decided to take up medicine as a profession. In 1649 he submitted his doctorate thesis at Leiden. In the same year he published his first work, a book on the geography and history of Japan, an excellent compilation in view of the limited material at his disposal. This was immediately followed by a companion volume on the religion of Japan. In August 1650 he published his *Geographia Generalis*, which was written between the autumn of 1649 and spring of 1650. The work should undoubtedly have been followed by a second volume, but it remained unfinished owing to his premature death, at the age of twenty-eight, in 1650.

Geography is defined by Varenius as follows: 'Geography is that part of *mixed mathematics* which explains the state of the earth and of its parts, depending on quantity, viz., its figure, place, magnitude, and motion, with the celestial appearances. By some it is taken, in too limited a sense, for a bare description of the several countries, and by others too extensively who, along with such a description, would have their political constitution.'

He divided geography into two divisions, general or universal, and special or chorography. The first he divided into three parts:

1. The Absolute part, which deals with the form, dimensions, and position of the earth; the distribution of land and water, mountains, woods, deserts; hydrography and the atmosphere.

2. The Relative part, which deals with the 'Appearances and Accidents that happen to it (the earth) from celestial causes', i.e. latitude, climatic zones, longitude, &c.

3. The Comparative part, which contains 'an explication of those properties which arise from comparing different parts of the Earth together'.

Varenius did not deal with Special Geography, but outlines its contents under three heads:

1. Celestial properties—the appearance of the heavens and climate.

2. Terrestrial properties, 'those which are observed on the face of every country', viz. position, boundaries, shape and size, mountains, rivers, woods and deserts, fertility, minerals, and animals.

3. Human properties, i.e. the description of inhabitants, their appearances, arts, commerce, culture, language, government, religion, cities and famous places, and famous men.

'These are the three kinds of occurrences to be explained in Special Geography, and though the last sort seem not so properly to belong to this science, yet we are obliged to admit them for custom sake and the information of the reader.'

Varenius complains that special is always taught at the expense of general geography, on account of which he argues that geography scarcely merits the dignity of a science. In special geography features should be explained in terms of general laws, so as to make local geography logical and intelligible.

Mathematical Geography is the best part of the book, since most data were available for this branch of the subject. Varenius deals with the rotundity of the earth, its dimensions and movements, and was the first to introduce in this connexion the works of Copernicus, Kepler, and Galileo into a geographical treatise. He discusses the divisions of the globe—the equator, tropics, and polar circles and 'climates' as known to the ancients, based on the length of the longest day (cp. Apian); latitude and longitude and methods of determining them, and methods of drawing map projections.

Meteorology. Varenius considers that the study of the air, its composition and physical properties, and the laws of motion, should form the basis of meteorology. Air consists of vapour and smoke exhaled from the globe, due to solar and its own heat. It has weight, and pressure is greatest near the earth's surface. It expands with heat and contracts with cold. He conceived of the air as expanding laterally, not vertically, and winds as due to this lateral displacement. The sun is the source of heat, therefore the movements of the air follow the sun, especially in tropical regions where the solar heat is greatest. Here, as the sun moves from east to west, so the winds (Trades) move mainly in the same direction. He gives a full account of the monsoon winds in the Indian Ocean. He cannot account for rain, but notes that precipitation is heavier on mountains due to rising vapours. He gives details of the climate of different regions. He notes two seasons in the Tropical Zone, winter and summer, one dry and one wet, and attempts to account for the climates of the areas he reviews (west coast of Africa, southern Arabia, the coast of Peru). He passes more rapidly over the Temperate Zone, though it is noted that northern China, though in the same latitude as Italy, has colder winters. For the polar regions, he gives a résumé of the voyages of Barents to Spitsbergen and Novaya Zemlya (1594–7).

Hydrography. The seas occupy cavities in the earth's crust, and they all communicate with each other. He states that the level of the Mediterranean is below that of the Atlantic, owing to the small strait which joins them; and places the Red Sea higher than the Mediterranean, though he advocates the construction of a canal to join them. Ocean movements are of two kinds, currents and tides. Currents he explains as due to the pressure exerted by winds on the ocean surface. He notes the east-west equatorial currents in the Pacific and Atlantic, and also notes a current issuing from the Gulf of Mexico between Cuba and Yucatan, i.e. the Gulf Stream, though he did not realize its climatic importance. He does not agree with Descartes that currents are due solely to the influence of the moon, but admits that they are strong at periods of new and full moon.

The source of rivers was always a great problem, and it is explained by Varenius by seepage from the seas, the water losing its salts by filtering, and rising up to the mountains by capillarity.

Physical Features. The greatest mountains have always existed, but others have been formed through accumulation by wind action. On the origin of mountains he writes that they are usually considered 'to have had a Being ever since the Creation', but 'since we can perceive a natural Decay and Corruption of them, we may judge they do not proceed from a supernatural origin'. That 'little mountains' have been built up by natural agencies is instanced by shell beds which occur in the hills of Gelderland; but 'large mountains' are probably 'of the same Age and Origin with the earth itself'. Yet he is doubtful of this idea, because theologists argued that the waters of the earth collected in channels, from the material of which the lands were built up. 'But we leave it to them to prove whether the Mountains be so many, and so large, as to fill all the channels of the sea.' Erosion by the sea, but not by running water, is admitted. Rivers carry material, deposit it by flooding their banks, filling estuaries and straightening the coast lines. The sea builds sand banks, and sea irruptions have flooded lands as in the Netherlands. All these facts reflect Varenius' first-hand knowledge of the north European plain, and particularly Holland.

Varenius, in the preparation of this work, was hindered by lack of material; the ideas lying behind the book were far in advance of the knowledge of his time. It remained the standard work for over a hundred years and was translated into Dutch, French, and German, and in 1672 its translation into English was arranged by Newton to be read by his students.

Some years before the publication of Varenius' book two other works were prepared, one by Nathaniel Carpenter (1625) and the other by Philip Cluverius (Cluver), a German (1624). Cluverius was born in 1580 at Danzig, and went to Leiden University to study law. He had no inclination for jurisprudence, and in 1601 a quarrel with his father ended his university career and he travelled to Bohemia. For two years he fought in Hungary against the Turks. He travelled widely in Europe during the period 1607 to 1613 and spent much time in London after 1610. In 1615 he returned to Leiden and published his *Germania Antiqua* in 1616 and was writing his *Italia et Sicilia Antiqua* up till the time of his death (1622). His general work, *Introduction to Universal Geography*, was not published till after his death (1624).

Cluverius approached geography through classics and history, and he was struck by the lack of descriptive material to serve as a basis for the full understanding of history. He therefore travelled extensively in order to produce his two works on Germany and Italy. In his *Universal Geography* he preserves the distinction between geography and chorography, but of the six books in the volume, only one deals with the earth in general, while the remaining five contain descriptions of countries, in which, as one would expect from his training, the human and historical elements are stressed. The first book of general geography is decidedly inferior to that of Varenius. He does not know of the views of Copernicus; for him, the earth is the centre of the universe. His mathematical geography and astronomy show no advance on the work of Apian written a century before. His physical geography is limited to the distribution of land and water.

But it is in regional description that Cluverius excels. The countries are described as follows:

1. Name, extent, and nature of the land and its products.

2. Ancient, and

3. Modern political divisions, ethnography, and topography. Volumes ii to iv deal with Europe in considerable detail; v and vi deal with the rest of the world more briefly owing to the lack of information.

Nathaniel Carpenter, who was elected a fellow of Exeter College, Oxford, in 1607, was the first Englishman to write a scientific geography. In his *Geography delineated forth in Two Bookes* he regards cosmography, geography, chorography, and topography as parts of a whole. He divides his book into two parts, spherical and topical: the first dealing with mathematical and physical geography, the second with the distribution and causes of phenomena treated on a general, not a regional, basis.

The first book is largely derived from the works of medieval and contemporary writers, Apianus and Gemma Frisius and Sacrobosco. Chapters on 'Partial Magnetical Affections' and 'Total Motions Magnetical' are obtained from the classical work of William Gilbert. The opening chapters of Book II deal with the habitability of the world, mathematical aspects of topography, 'the manner of Expression and Description of Regions' and 'the use of instruments for finding out the Position of two places'. Chapters V to VIII deal with hydrography. The distribution of seas is described as 'due either to the collection of the waters which formerly thinly covered the whole earth, into smaller areas, thus leaving dry and habitable land; or that God, in the creation, made cavities in the Earth in which the waters were collected and confined'. Chapter VI deals with currents, tides, and winds, Chapter VII with the 'Depth, Situation, and Termination of the Sea', and Chapter VIII with sea traffic and merchandise. Then follows an account of the land and its features. He discusses the value of river sites for towns thus: 'first, because by means of such water they have quick passage and traffique . . ., second such a site is most convenient for purging away filth . . ., thirdly because such rivers yield store of fish . . ., fourthly [water would be near in case of fire] . . ., lastly amongst other reasons we cannot forget the pleasantness of faire rivers. . . .' In discussing 'Mountaines, Valleyes, Plaine Regions, Woods and Champain Countreys' it is stated that 'Mountains since the beginning of the world have still decreased in their quantity, and so will continually decrease until the end' and 'the causes we shall find to be the Water, as well of Rain as Rivers'. He thus recognizes the efficacy of rain and river erosion. Carpenter also discusses the character of sea cliffs and their formation, and ascribes similar precipitous rock faces which occur inland either as 'by nature fashioned craggie and uneven' or to earthquakes. The remaining chapters deal with the 'civil affections of the land', that is, 'those which concerne the Inhabitants'. A large part is taken from the works of the French savant, Jean Bodin, who had developed fully the idea of the relations between physical conditions and human progress. The inhabitants of the world are divided into northern, middle, and southern types, each type related to its environment, in physical and mental characteristics and natural aptitudes. Thus 'mountain people are for the most part more stout, warlike, and generous than those of the plaine countries, yet less tractable to government'. It is argued that a change of environment results in a change of manner, for 'colonies transplanted from one region into another farre remote, retaine a long time their first disposition, though by little and little they decline and suffer alteration'. But Carpenter does not unreservedly accept these borrowed conclusions, for he asserts that man is not solely influenced by environment.

Carpenter's work was eclipsed by those of Cluverius, who set the standard in regional geography, and Varenius for general geography; regional descriptions are not included in his treatise, whereas his mathematical and physical geography, though prepared with caution, and displaying common sense and judgement in the selection of authorities, is inferior to the work of Varenius. His work was neglected, and Cluverius and Varenius were the standards for over a hundred years.

Chapter XIII

MEASUREMENT AND CARTOGRAPHY (1650–1800)

ESSENTIAL to navigation and the construction of accurate maps are methods and instruments for the precise determination of position—i.e. latitude and longitude. Latitude may be determined by measurement of the altitude of the sun, the pole star, or the upper and lower culmination of a circumpolar star. The earliest instrument for measuring the elevation of the sun is the gnomon, which consists simply of a vertical rod, from the length of whose shadow the altitude of the sun can be calculated. This, as we saw in Chapter III, is said to have been improved upon by Aristarchus, in the form of the scaph, consisting of an upright rod, whose shadow could be measured in a bowl. The astrolabe, which dates back at least to Hipparchus, and in principle probably earlier, is the forerunner of the modern sextant. It was used by mariners till the seventeenth century for the direct measurement of the altitude of the sun. The quadrant worked on the same principle, but it was less cumbersome, as it only consists of a quarter of a circle. The cross staff, again serving the same purpose, is first mentioned in 1342, and a modification of it—the back staff, which used the reflected rays of the sun—was invented by John Davies in 1585. For the accurate determination of latitude, the first ephemerides were prepared and published by Regiomontanus for 1474–1506; Peter Apian also made a series for 1521–70, but the results were not accurate until the compilation of Kepler's Rudolphine Tables in 1526.

The determination of longitude was much more difficult, until the eighteenth century, for two reasons; there was first the problem of choosing a prime meridian, secondly the difficulty of calculating the angular distance east and west of this line. Ptolemy had used the meridian of the Fortunate

Isles, vaguely identified with the Canaries, as his standard. In late editions of Ptolemy, the Azores and Canaries are shown in the same latitude, whereas in fact they are 18° apart. Columbus was the first to suggest that the amount of declination of the needle would give the longitude of a ship,

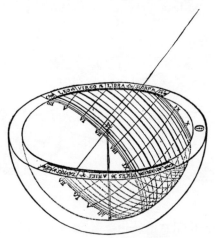

FIG. 18. The Scaph.

and in 1530 was produced the first crude map to show magnetic variation; but it was soon realized that the method was unreliable owing to the annual magnetic variation. The magnetic meridian lay to the west of the Canaries, and Mercator sought to select this as his prime meridian, ultimately deciding upon the Azores in his 1569 map—though in his first world map he had used Ptolemy's meridian. The Spanish frequently used the Tordesillas Line, 370 leagues west of the Cape Verd Islands—this, for instance, is the prime meridian on the first map of the New World prepared by Juan de la Cosa. There was no unanimity in the selection of the standard meridian till the eighteenth century. Richelieu decided on Ferro, in the Canaries, as the standard for French maps; while Sir George Fordham has suggested that the meridian of London was first used as a standard in the late seventeenth century, in an English county map.

Longitude was obtained by observing the occultation of stars by the moon in the sixteenth century and after, a method popularized by Jean Schoner and Peter Apian. But even

FIG. 19. Cross-staff, 1594.

to-day this method is inaccurate, and 'geographical longitude continued to be a torment to science. Fine theories were elaborated, admirable propositions put forward, and each time heaven and nature confounded the scientists' (Lelewel). In 1611 Galileo invented his telescope, which permitted the more accurate observation of celestial phenomena, and in 1611 he observed the satellites of Jupiter, and suggested a method of determining longitude from their occultation, which was later perfected by Laplace. But high-power telescopes could not be used on moving vessels. The fundamental difficulty was due to the need for accurate time-keepers, as realized by Frisius in 1530, to permit the comparison of the time of occurrence of similar celestial phenomena, and the measurement of longitude was not rendered simple or accurate till the invention of John Harrison's chronometer in 1735.

In 1666 the French Academy of Sciences was founded and for the next hundred years it led the way in the development of astronomy, geodesy, and cartography. Jean Fernel (1497–1558), as far back as 1528, had made a remarkably accurate measurement of the length of a degree, by counting the

FIG. 20. Davis's Back-staff, 1594.

revolutions of a carriage wheel between Paris and Amiens. In 1669 Jean Picard made the first geodetic measurements between the same places. He used a quadrant fitted with a telescope, with crossed wires to ensure accuracy. In 1657 Charles Huygens, a Dutch scientist, invented the pendulum clock (later perfected by Harrison for use on ships), and Jean Richer, using one for astronomical work in South America on behalf of the Academy of Sciences, found that the pendulum regulated to beat seconds in Paris failed to do so at Cayenne, 5° from the Equator. This agreed with the mathematical speculation of Newton, who argued that owing to centrifugal force, due to the rotation of the earth, there would

be a flattening of the earth at the poles and a bulging at the
Equator. To substantiate this hypothesis as against the con-
flicting evidence obtained by J. and D. Cassini in observations
in France (which were subsequently proved to be inaccurate),
the Academy dispatched expeditions to Peru and Lappland to
measure arcs for comparison. The decision was given in
favour of Newton by the Academy—'Thus', to quote Voltaire,
'[they] flattened both the Poles and the Cassinis.'

The Academy of Sciences dispatched expeditions throughout the world to obtain accurate astronomical observations of
longitude and latitude. Longitudes for France were collected
by Picard and Lahire in 1672–80, and were subsequently
used for a map of France prepared by D. Cassini (1694).
Many of these measurements were published in the *Con-
naissance des Temps* in 1679.

In 1662 the Royal Society of London was founded, and
it also instigated much scientific work, which included the
voyage of Cook to investigate the Transit of Venus, and that
of Captain Phipps (1773) to the Arctic to see how far north
navigation was practicable. In 1675 the Greenwich Observatory was founded, and a Board of Longitude was established
in 1713, the Commissioners of which conceived the plan of
a Nautical Almanac, which was first published, under the
auspices of the Astronomer Royal, in 1767.

Methods of survey rapidly improved after 1500 and many
accurate maps of relatively small areas were produced. The
forerunner of the prismatic compass was an adaptation of the
mariner's compass, the needle being mounted on a graduated
flat disk. Such an instrument was used at the end of the
fifteenth century in Italy for surveying, and was also used by
Waldseemüller and Gemma Frisius. Frisius was the first to
describe the method of fixing position by intersecting rays, and
the method was elaborated, as noted above, by Munster; while
further, with the use of the compass, a detached trigonometrical survey of a small area was prepared by an Englishman,

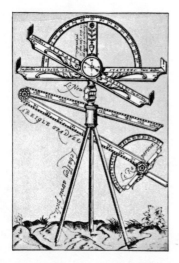

Plate V. (*a*) Graphometer by Philip Danfrie, 1597, using a
fixed and a movable alidade, the observed angle between them
being drawn on the plan either directly or by using the hinged
rule illustrated behind the graphometer

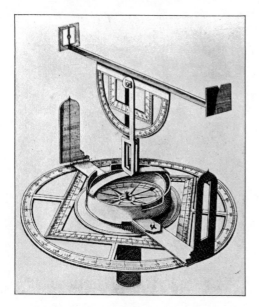

(*b*) An early theodolite, represented in Blaeu's Atlas, 1664

William Bourne, at the end of the sixteenth century. Most of the detailed accurate maps of different countries in the sixteenth century were based on the fixation of the position of a few places astronomically, and then the addition of detail with the compass. Modern methods of surveying were definitely established by a Dutchman, William Snell, in 1617,

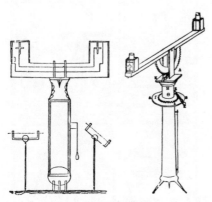

FIG. 21. The Diopter.

for he measured an arc between Alkmaar and Bergen-op-Zoom by means of triangulation and trigonometrical calculation. In 1571 Leonard Digges introduced to England a crude form of theodolite, based in principle on the 'diopter', and used for range-finding. It was later perfected by Jonathan Sission (1737), and some years later by Ramsden, whose instrument was used for the survey of England and Wales, begun in 1784. At the end of the century the plane table was invented by Jean Pretorius of Wittenberg (1590) and Philip Dandrie, a Frenchman (1597).

During the seventeenth and eighteenth centuries, maps based upon astronomical observations and careful surveys were produced for most of the countries in western Europe. A high standard in cartography was set by the *Carte de la*

France prepared by J. and D. Cassini in 1694, based upon their trigonometrical survey and the observations of Picard and Lahire. In the middle of the eighteenth century Cassini de Thury, at the instigation of Colbert, corrected, and used

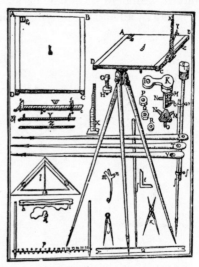

FIG. 22. Pretorius's Plane-table, 1594.

as a basis for a detailed survey, the observations and triangulation of J. and D. Cassini, and was responsible for establishing a geodetic connexion between Paris and Greenwich (1784). The topographic maps of France were prepared slowly till the Revolution, when the Government took the survey in hand and quickly completed it (scale 1/86,400).

In Britain Christopher Saxton produced the first atlas of England and Wales (1574-9) in which the maps are on an approximate scale of one inch to three miles. This was followed by Timothy Ponts's maps of Scotland (1608) and John Speed's in his *Theatre of the Empire of Great Britain*, on the same scale as Saxton's in 1610. Hollar, using a scale of one inch to some five miles, prepared maps of England and Wales in 1644, which were of service during the Civil War. Later, as a response to the need of accurate survey for military requirements, a survey of Ireland was begun in 1653 to facilitate the allocation of property to those who assisted in the suppression of the Irish rebellion. Other noted English cartographers and geographers of the period, whose works cannot be more fully mentioned here, were John Ogilby, who was the first to adopt the modern linear measure of 1,760 yards to one mile as a substitute for the former scale of 2,420 yards, and prepared careful road maps in his *Britannia* (1675), the first reliable work of its kind; John Cary (*The New and Correct English Atlas, being a new set of County Maps from actual Surveys*, 1793), Emanuel Bowen, Richard Blome, and Robert Morden, who were all responsible for good county surveys. In 1747, after the Scottish rebellion, Captain (later General) Roy began a survey which proved to be the birth of the Ordnance Survey. Roy supervised the English share of the geodetic connexion between Paris and Greenwich already mentioned, and so began our first national triangulation. In 1791 the preparation of a large-scale topographic map was started for military purposes, and the Ordnance Survey was established for the purpose.

J. F. W. Desbarres used the nautical surveys of James Cook in his work on the Atlantic (*Atlantic Neptune*, 1774). Thomas Jefferys prepared West Indian and American atlases. Major James Rennell, when surveyor-general of Bengal (1763), mapped the Ganges and Brahmaputra, and surveyed Bengal and Bihar (1763–82) on a scale of five miles to one inch, and published a map of India in 1788. He also spent many years collecting log-books to investigate the currents of the North Atlantic.

In the field of world cartography the laurels in the eighteenth century go to two Frenchmen, Guillaume Delisle (1675–1726) and J. B. B. D'Anville (1697–1782). All the new and old data were collected and corrected by Delisle, and he produced a map of the world (1700) which was the most accurate map yet compiled, entirely free at last of Ptolemaic

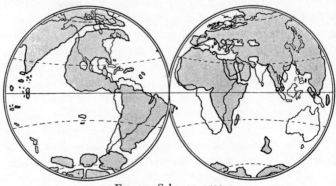

FIG. 23. Schoner, 1523.

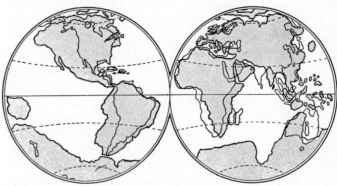

FIG. 24. Mercator, 1587.

FIGS. 23–6. Comparative View of Outlines of the World, 1523–1761.

tradition, with the Mediterranean given its correct length of 41° longitude, and with Ferro, 20° W. of Paris, as the standard meridian. But the brilliant work of Delisle was eclipsed by that of his successor D'Anville, who produced in all some two hundred maps, including an *Atlas General* (1737–80). His world map is almost perfect in its outline of the land masses, for he had more records than Delisle. But D'Anville's

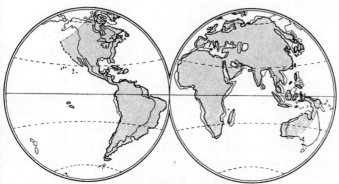

Fig. 25. Homan, 1716.

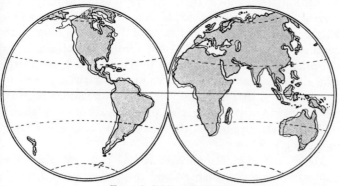

Fig. 26. D'Anville, 1761.

great contribution to cartography was his rigorous exclusion of all features which were not proven; his cartography was truly scientific. The imaginary lakes, the fantastic rivers, and the Mountains of the Moon, a Ptolemaic inheritance, were excluded from Africa, and its interior was left blank. He mapped China on the basis of a survey undertaken at the instigation of a Chinese emperor by Jesuit missionaries in 1718, and he wiped out the mythical continent from the south seas.

Methods of cartographic representation also made great strides in the eighteenth century, particularly those of showing relief. The old perspective method was still generally used (as in the maps of Delisle and D'Anville) but crude hachures are first used on some of the French topographic maps noted above, the earliest about 1674, on a map of the environs of Paris by David Vivier, and also on Cassini's *Carte de France* (1694). M. S. Cruquius first used contours in his chart of the Merwede in 1728, and P. Buache used the same method to show the depths of the English Channel in 1737. A contoured map of France was produced in 1791 by Dupain-Triel. The contour and hachure methods were first scientifically combined by J. G. Lehmann in 1783; and thereafter this was the principal method of showing relief on all the national topographic surveys during the nineteenth century. The development of hill-shading is well shown in the atlas of Germany prepared by Homan of Nuremberg, who also produced a very accurate map of the world in 1716, comparable to that of Delisle and based on similar records. The first map of Homan's atlas, prepared about 1718, shows the earliest attempt to vary the intensity of shading according to the degree of slope.

Meanwhile, the accumulated data of two centuries were gradually being systematized and mapped to serve later as basis for the development of geographical synthesis. The variation of the compass was mapped by Christopher Burrus in the early seventeenth century and was greatly improved on by F. Halley in 1683. Halley, 'the father of dynamical meteorology', in 1686 produced his wind chart, and enunciated his epoch-making theory of the Trade-winds (1696), in which he first attempted to relate the general circulation of the atmosphere with the distribution of solar heat on the earth's surface. George Hadley was the first to give the correct explanation for the deflexion of the trade-winds by the earth's rotation (1735) which was elaborated a hundred years later by John Dalton (1834). A. Kircher mapped, as far as his records would allow, the currents and greatest depths of the oceans in 1665 in his *Mundus Subterraneus*, and Rennell and Alexander Dalrymple, who had collected many records, produced nautical memoirs and charts in the second half of the eighteenth century.

The synthesis of the vast body of new facts relevant to physical geography was attempted at the close of the eighteenth century by such men as Buache (*Essai de géographie physique*, 1756), Torbern Bergmann, a Swedish chemist (*Physical Description of the Earth*, 1766, which was translated into English, 1772, and German, 1774), J. R. Foster in 1783, and lastly Emanuel Kant. Bergmann (1735–84) held theories regarding mineral structure and the constitution of the earth's crust, which were largely adopted by Werner, the German geologist, who exercised such great influence on scientific thought in the early nineteenth century.

Kant lectured on physical geography in the University of Königsberg from 1765 onwards, and his lectures were subsequently published. In his view the human element was an integral part of the subject matter of geography. 'Geography appealed to him as a valuable educational discipline, the joint foundation with anthropology of that knowledge of the world which was the result of reason and experience' (article on Geography, *Encycl. Brit.*). The communications of experience from one person to another he divided into two branches, narrative or historical, and descriptive or geographical, but both history and geography he regarded as descriptions, the former in time and the latter in space. Physical geography, which he claimed to be 'a summary of nature, is the basis of history and all the other possible geographies', of which he names five—mathematical geography (the form, size, and movements of the earth and its place in the solar system), moral geography (the customs and character of man in relation to environment), political, mercantile (commercial), and theological geography (the distribution of religions).

The state of geographical science on the eve of the scientific development of the nineteenth century is well summarized by Pinkerton's *Modern Geography* published in 1807. Geography is explained 'as embracing topics of the most multifarious description', and current geographies are deplored as 'arranged in the most tableless manner and exceeding in dry names and trifling details . . . .' To the writer, geography, 'like chronology, only aspires to illustrate history'; and it is the business of the geographer, like that of the architect, 'to erect a solid and elegant edifice from materials already prepared'. Pinkerton gives full references to all his authorities, among which his particular indebtedness to French scientists is a notable feature. He maintains the distinction between geography, chorography, and topography. The popular view of geography, he states, is the 'description of the various regions of the globe, chiefly as being divided among various nations and improved by human art and industry'. A more suitable definition, he argues, is historical geography, which he divides into the geography of ancient or classical times, the middle ages, and modern geography. 'The chief object of modern geography is to present the most recent and authentic information concerning the numerous nations and states who divide and diversify the earth.'

With regard to the distribution of land and water, he recognizes two types of divisions, terrene or continental, and maritime. There are two terrene divisions, the old world and the new. Notasia (usually called 'New Holland' with the adjacent islands, i.e. Australia) and Polynesia, form the two maritime divisions. The divisions, or land forms, to use the modern term, are of the simplest kind, and show no advance on the early sixteenth-century conceptions—bays, straits, gulfs, rivers, continents, islands, peninsulas, and isthmuses. The distribution and arrangement of the physical features of the earth to the uninitiated appear irregular and fortuitous, like 'an immense ruin', but they are, in fact, 'most beneficial and even necessary to the welfare of its inhabitants, for to say nothing of the advantages of trade and commerce, which could not exist without seas . . . it is only by their vicinity that the cold of high latitudes is moderated and the heat of the lower'. The current conception of the form of the lands was that which was established by Buache. It is the conception of basins or concavities, from the oceans down to small river basins, each bounded by mountain ranges. This view Pinkerton does not unreservedly accept, for rivers do not always occupy definite basins, and they often cut through gorges in mountains, 'so estranged is nature from human theory'. Thus 'the theory of the French geographers, though just in general, must not be too widely accepted'.

He follows Halley closely in describing the winds. These he describes and divides into three groups, variable, constant (the trades), and seasonal (monsoons). The eastward deflexion of the trades he, like Varenius, and apparently not knowing Hadley's work, ascribes to the movement of the sun, and the consequent westerly shift of the point of greatest heat, the air drawn in thereby being compelled to rush in from the east, and so form a constant easterly wind. The limit of the trade wind belt, accompanied by winds in the opposite direction in the upper air, though admitted as variable, is placed correctly at about lat. 30°, and the belt of calms at the point of greatest solar heat. The monsoon effect on the coast of Guinea was noted by Halley, and the monsoons are ascribed to the rapid heating of a land mass adjacent to a sea. But in the Indian Ocean, Pinkerton is puzzled why 'this change of the monsoons should be any more in this ocean than in the same latitudes in the Ethiopic where there is nothing more certain than a south-east wind all the year'. Again reflecting Halley's researches, weather conditions are associated with variations of barometric pressure. Moreover, it is concluded that the daily variations of pressure in the tropics are slight as compared with the temperate zone. 'Hence I conceive that the principal cause of the rise and fall of mercury is from the variable winds which are found in the temperate zone. . . .'

With regard to temperature, he estimates the average annual temperature in different latitudes over the Atlantic and Pacific, where, owing to the absence of land, the temperatures must be most near to the normal. Thus he concludes that the temperature variations from year to year are small near the Equator and increase to the Poles; it scarcely ever freezes inside lat. 35°, or hails beyond 60°; and between 35° and 60° near the sea, thaw sets in when the temperature is about 40° F. The smaller temperature range over the seas than on the land is noted. January is usually the coldest month, July the hottest above 48°, and in lower latitudes, August. The effects of currents on ocean temperatures is discussed and the height of the snow line is given for different latitudes. An important observation is that every habitable land has a temperature of 60° F. at least for two months, and this is stated to be necessary for the ripening of corn.

In his description of countries his exaggerated historical conception is apparent in both the treatment and arrangement. The order of description for each country is as follows, and is the same arrangement as that adopted by Robert de Vaugondy in his *Essai sur l'histoire de géographie*:

1. Historical or progressive geography of each country.
2. Political state, including statistics.
3. Civil geography.
4. Natural geography.

While some consider that the fourth should come first,

Pinkerton claims that the landscape is so much the product of human industry that the above order is correct.

From the foregoing it will be obvious that from the standpoint of modern geography, on its physical side, development is proceeding on scientific lines with the accumulation of data and the progress of research. But the essence of the modern concept, the nature of the adjustment of man's activities to the physical environment, is still a scientific aspect foreign to the eighteenth century. The nineteenth century was to witness the collection of relevant data by explorers, arrangement, synthesis, and correlation by the scientists, and the gradual refinement of the concept of the evolution of the mutual interdependence of man and environment. These advances were due in the main, in the first half of the nineteenth century, to Alexander von Humboldt and Karl Ritter, the founders of modern physical and human geography. But before studying their contributions, the progress of exploration and cartography—that is the accumulation of data and their representation—will be briefly summarized.

Figs. 18–22 are reproduced from R. T. Gunther, *Early Science in Oxford*, 2 vols., by permission of the author.

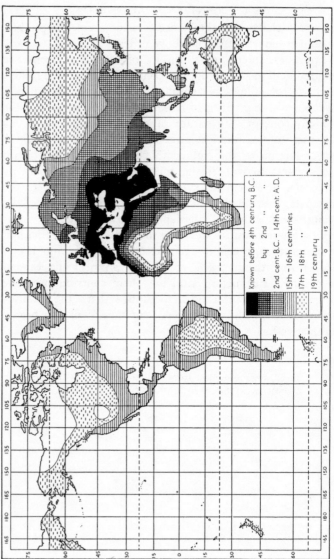

FIG. 27. The Progress of Exploration.

Known before 4th century B.C.
" by 2nd " "
2nd cent. B.C. – 14th cent. A.D.
15th – 16th centuries
17th – 18th "
19th century

Chapter XIV

EXPLORATION AND CARTOGRAPHY IN THE NINETEENTH CENTURY

THE two great drawbacks to the scientific development of general mathematical and physical geography, and the systematic co-ordination and interrelation of facts of physical environment and human activities and organization, were firstly, the lack of accurate data, the limited knowledge of the known lands, and ignorance of the general distribution of land and water; and secondly, the need for accurate cartographic representation of the distribution of physical phenomena over the earth and the evidence of man's occupation on its surface.

The foundation of a truly scientific treatment of the surface phenomena of the earth was therefore attendant upon the progress of scientific exploration and cartography and the co-ordination of data in the natural and physical sciences. The nineteenth century, in these spheres, is characterized firstly by the exploration of the continental interiors—as opposed to mainly coastal exploration of the sixteenth to the middle of the eighteenth century—by explorers as intent upon the collection of new facts about the earth's surface, as upon the discovery of new physical features to be placed upon blanks on the world map. Secondly, it is distinguished by the development of national large-scale topographic surveys. Before discussing the development of the science of geography in the nineteenth century the main features of the development in these two directions will be briefly summarized.

Exploration. After a long period of stagnation during the eighteenth century, the active exploration of the interior of Africa may be said to begin with the inauguration of the African Association in 1788. Before this date the only important expedition was that of James Bruce who visited

Abyssinia in 1770, discovered the source of the Blue Nile, and traced it to its junction with the White Nile at Khartoum. In 1800 the courses of the great rivers were still unknown, and Ptolemy's Mountains of the Moon still figured on maps of Africa as one of its principal physical features.

In 1795 the African Association sent out Mungo Park, a Scotsman, to explore the Niger. He travelled up the Gambia (which had previously been considered to be the mouth of the Niger), reached the latter river at Sebu, and then traced its course for three hundred miles. On his second journey in 1805, after tracing the course of the river for a thousand miles, almost to its mouth (which he considered to be that of the Congo), Mungo Park and all who were with him, except one guide, perished. In 1823 W. Oudney and H. Clapperton reached Lake Chad, and Clapperton afterwards explored the strongholds of the Bornu and Hausa civilizations. In 1830, after the death of his master Clapperton, Richard Lander solved the problem of the outlet of the Niger by starting from the Guinea coast and following its course from Bussa to its mouth in canoes. In succeeding years Timbuktu was visited for the first time by Europeans in 1826 by A. G. Laing and in 1828 by R. Caillié. H. Barth, who had already travelled widely in northern Africa, ascended the Nile to Wadi Halfa, and set out in 1850 on a trading mission on behalf of the British Government to central Africa. He travelled on this occasion from Lake Chad to Timbuktu, and studied in detail the native civilizations with which he came into contact.

The two great problems for African exploration after 1850 were the sources of the Nile and the Congo, and with their solution must always be associated the names of R. F. Burton and T. H. Speke, David Livingstone and H. M. Stanley. Abyssinia and the chief tributaries of the Nile were fairly well known by 1850, but above Gondokoro, owing to the seventy miles of rapids and the opposition of the fierce Bari tribes, its upper reaches were still unknown. Now the attack was made from a new direction instead of up the river. Jesuit missionaries, finding the natives more tractable than in Abyssinia, made their head-quarters at Zanzibar, whence they discovered and made known the snow-covered mountains and lakes of the equatorial region of east Africa, believing the lakes to be part of a great inland sea. After a first expedition from Aden into Abyssinia and Somaliland, Burton and Speke set out from Zanzibar in 1856 to find the ultimate source of the Nile. They reached the plateau of Unyamwezi and travelled thence to Tanganyika. From here Speke proceeded alone owing to the illness of his partner, and he saw from a distance Victoria Nyanza, which he was convinced formed the source of the Nile. In 1860 Speke set out again to extend his discoveries, this time in company with J. A. Grant. He made a rapid survey of Victoria Nyanza, explored the unknown country of Uganda, and found the Ripon Falls where the Victoria Nile leaves the lake; thus he solved the problem of the source of this river. Samuel Baker, who met Speke and Grant at Gondokoro, in the face of great dangers and difficulties discovered (and incidentally greatly overestimated the size of) Albert Nyanza, and travelled along the Nile as far as the Murchison Falls. Subsequently the Nile and its tributaries were surveyed by English officers, and G. Rohlfs and G. Nachtigal made extensive explorations in the Sahara and the Sudan (1860–75).

Meanwhile the veil was gradually being lifted from the other great problem, the source of the Congo. Livingstone, missionary and explorer, had been in Africa since 1840, but it was not till 1849 that he made his first great journey across the Kalahari to Lake Ngami. Then in 1852 he travelled up the Zambezi, discovered the Victoria Falls, and crossed to the west coast—the first journey across equatorial Africa. His last and greatest journey, which aimed at the discovery of the watershed between Lakes Nyasa and Tanganyika, was begun in 1866. He set out from Zanzibar, and, despite serious illness, journeyed from Lake Nyasa to Tanganyika, discovering lakes Mweru, Mofwa, and Bangweulu, and the river Lualaba, which he thought was the upper Nile. At Ujiji he was met by H. M. Stanley (1871) who brought assistance, but in 1873 Livingstone died. His work was followed up by Stanley, who undertook one of the most remarkable journeys in the history of African exploration (1874–7). After further exploring the East African lakes, he crossed Africa from east to west from Zanzibar, and travelled down the Lualaba, which he proved to be a tributary of the Congo. It was this journey, opening up hitherto unknown country, which led to the formation of the Congo Free State. In 1884, supplementing Stanley's work, Joseph Thomson explored Mounts Kilimanjaro and Kenya; and Wissmann and other German explorers (1881–6) explored the southern tributaries of the Congo, Wissmann himself crossing the continent from west to east in 1881–2. Stanley in his last expedition (1887–9) opened up the forest of central Africa, extended our knowledge of Lake Albert Edward, and discovered the snow range of Ruwenzori which was later ascended and mapped by the Duke of Abruzzi (1906).

Scientific work in the early nineteenth century in Asia was mainly confined to India, with the initiation of its great trigonometrical survey in 1800. The scientific exploration of the plateaux bordering India was left till the second half of the century and after. Indian native surveyors in the late nineteenth century were able to penetrate to Tibet and even to Lhasa (1863–82), though the sacred city was closed to Europeans. Thomas Manning, in disguise, at the beginning of the nineteenth century, and the Abbé Huc (1814) were the only two Europeans to enter Lhasa in the nineteenth century. The last quarter of the nineteenth century, however, witnessed the extensive exploration of Tibet from the Indian side, and by Russians from the north. The most famous of the latter was Nicolai Prjevalsky, who in 1871–3 and 1876 explored and surveyed the Tsaidam region, and in 1879 studied historic climatic changes in central Asia. Russians also were responsible for the exploration of the Aral and Caspian depressions. But the greatest contributions to the geography of central Asia have been made during relatively recent years by Sven Hedin and Sir M. Aurel Stein. Finally, China and its contiguous regions were widely explored by Richthofen (1868–72).

At the end of the eighteenth century the shores of Australia were thoroughly explored. George Bass explored the coast of New South Wales, and in 1797–8 discovered the strait now bearing his name which separates Tasmania from the mainland, and circumnavigated the island. Matthew Flinders (1801–3) explored the coast of Australia from King George Sound, round the south, east, and north-east as far as Arnhem Bay. The substitution of the name Australia for New Holland was due to the suggestion of Flinders. The early settlement of Sydney was cut off from the interior by the Blue Mountains, and after many fruitless attacks up their bottle-necked valleys, they were finally crossed by Gregory Blaxland in 1813. In his wake there followed a succession of explorers, bringing back with them reports of a vast inland sea, or a desert with dwindled streams, according to the season of their journeys across the plains. John Oxley, Richard Cunningham, Hamilton Hume (who hailed Lake George as part of an inland sea), W. H. Hovell, Charles Sturt (who discovered the mouth of the Murray in 1830), and Sir Thomas Mitchell (who set out fully equipped for the navigation of the hypothetical sea) all made their contributions to the opening up of the south-eastern interior of the continent in the first half of the century. F. W. L. Leichhardt and E. B. Kennedy opened up the north-east, E. J. Eyre and A. C. Gregory the south and west, while the Gregories (1856–62) found pastoral areas to the north and south of the western desert. Thus by 1860 all humid Australia had been explored. During the 'seventies the chief journeys were transcontinental, and no less than five expeditions were made across the central desert, by P. Egerton Warburton (1873), J. Forrest (1874), Ernest Giles (1875), and A. Forrest (1879). In 1860 John M'Douall Stuart from Adelaide, and Robert O'Hara Burke and W. J. Wills from Melbourne, set out to cross the continent from south to north. Both succeeded, but Burke and Wills perished during the expedition, while Stuart's route resulted in the discovery of the Macdonnell Ranges and was later followed by the overland telegraph (1872).

Between 1773 and 1779 the British Government sent three expeditions to the Arctic, led by Phipps (who reached 80° 37' N., where he was stopped by ice, to the north-west of Spitsbergen), James Cook to search for a north-east or north-west passage by way of the Bering Strait, and Clarke, who continued the same expedition after the death of Cook at Hawaii, passed through the Bering Strait, and reached 70° N. With the stimulus of large monetary rewards, Arctic exploration from the United States recommenced in 1815, led by J. Franklin, J. C. Ross, and E. Parry. During this period the archipelago to the north of America, and the northern coasts of this continent, were almost fully explored, and the northern coast of Siberia was surveyed by Russians.

In the north-eastern Arctic regions, the most notable journey was that of Lieut. Julius Payer, an Austrian, and Lieut. Weyprecht, who in a search for the north-east passage (1871) were beset by ice off Novaya Zemlya, and drifted till they reached Franz Jozef Land, later visited and more thoroughly explored by an Englishman, Leigh Smith. The north-east passage was ultimately made in 1879 by A. E. Nordenskiold. He was followed by Captain Joseph Wiggins, who voyaged in the Siberian seas, and these two last explorers proved jointly that the route as far as the mouth of the

Yenesei was a practicable commercial route. Greenland was crossed for the first time by Fridtjof Nansen (1888), and Robert E. Peary also made memorable journeys in the same country (1886–95). Later the Danes charted the east coast of Greenland, and the north-east coast was accurately surveyed for the first time by L. M. Erichsen (1905–7).

The polar area itself was made the primary objective of exploration, as opposed to the passages round the north of the two continents, after 1817. Most of such expeditions started from Spitsbergen, but Nansen, following the drift of the ice, passed from the New Siberian Islands to Spitsbergen in three years, though he reached no farther north than 85° 55′ N. Then he made an attack from Spitsbergen and reached 86° 5′ N. In 1899 Abruzzi reached 86° 34′ N. Nansen's journeys proved for the first time that the Arctic region consisted of a sea, the depth of which increased towards the Pole. The journey of De Long (1879), in which the whole party perished, proved the fallacy of the conception of a continent to the north of Bering Strait, stretching across the Pole to Greenland, and showed that, in fact, to the north of Siberia there was an ocean with many small islands. In 1909 Peary was acclaimed as having reached the Pole, but some now hold that he was mistaken, or that he was preceded there by Dr. F. A. Cook in 1908. Peary saw no land and the depth of the sea was recorded as 1,500 fathoms.

After Capt. Cook, who was the first to cross the Antarctic circle, there was little South Polar exploration till the voyage of the Russian, Fabian von Bellinghausen, in 1819. He sailed half-way round the Antarctic circle, voyaging within for considerable distances. James Weddell reached 74° 15′ S. in 1823, and he was followed by John Biscoe, an English sealer, in 1835, who sighted Enderby Land, the Biscoe Islands, and Graham Land, which lies behind those islands. In the 'thirties three expeditions were prepared in England, France, and the United States, to pursue magnetic observations in

the Antarctic, and to explore the extent of the southern
continent. Dumont d'Urville, the first to start, from France,
discovered Adélie Land. Charles Wilkes (America) was not
greatly successful owing to quarrels among his officers and
unseaworthy ships. The British expedition under Sir James
Ross started last (1841) and was much better equipped
for Antarctic exploration. Ross reached Victoria Land,
discovered the twin volcanoes Erebus and Terror (named
after his two ships), and then pushed south as far as 78° 4′ S.,
the most southerly latitude yet reached. Until 1898 there
were no more organized expeditions to the Antarctic, and
the only ships to cross the circle were those of sealers and
whalers. But in 1898 there were three further expeditions,
which set out with the South Pole as their objective. A
Belgian expedition reached Graham Land; a German vessel
rediscovered Bouvet Island; and an English expedition under
Borchgrevink, on the 'Southern Cross', wintered in the
Antarctic, reached Mount Terror, and sailed along the great
ice barrier as far south as 78°. In 1901 the attack was
renewed by R. F. Scott, and for the first time he approached
by land. He travelled along the ice barrier, and discovered
and named King Edward Island. He proved Mts. Erebus
and Terror to be on an island, and wintered there. Then in
the next summer he pushed south to 82° 17′. Meanwhile
a German party had discovered Kaiser Wilhelm II Land,
and two other expeditions were in Antarctic waters at the
same time. The numerous synchronous records, meteorological and magnetic, thus formed a valuable contribution
to science. Between 1902 and 1904 King Edward's Land,
Kaiser Wilhelm Land, Coats Land, and Loubet Land were
discovered, situated at the four quarters of the Antarctic
Circle. Sir E. H. Shackleton landed in 1908-9 at Mount
Erebus, and nearly reached the Pole—actually, 88° 23′ S.,
162° E. In January 1912 Captain Robert Falcon Scott with
four companions reached the Pole, but found that the

Norwegian Roald Amundsen (who previously had designed to make an attempt upon the North Pole) had preceded him there in December 1911. Scott and his party died on the return journey.

These expeditions have finally solved the general problem of the outline and extent of the southern continent. The mythical land-mass has been progressively reduced in fact by the progress of exploration southwards, till to-day it is known that the land-mass of the south is entirely within polar latitudes. Yet many scientific problems, appertaining to its extent, geology, physiography, climate, flora, and fauna, await detailed investigation. These have been the objectives of several expeditions by sea, land, and air, to both the north and south polar regions, in recent years.

Cartography. In the early nineteenth century (1818) a new map of France was begun on a scale of 1/80,000, based upon a new survey. It was completed in 1878. This map is engraved, all in black, with relief shown by hachures and spot heights. Fruitless efforts were made during the nine-teenth century to undertake a large-scale survey of the whole of France for the production of a new standard map on a scale of 1/50,000 to supersede the existing map (1/80,000). It was not till 1898 that such a scheme was adopted by the Service Géographique de l'Armée. The new survey is on a scale of 1/20,000 and the 1/50,000 map is reduced from these base sheets. At present only about one-seventh of the surface area is surveyed on the large scale, and the 1/50,000 is only published for the neighbourhood of Paris, Lorraine, and part of the coast of Provence, one-thirteenth of the total surface area. The first edition of the 1/50,000 was an elaborate and expensive lithographic map with eight colours, different shades of green for forests, parks, and meadows; violet for vineyards; red for towns, and three colours for relief shown by contours in addition to vertical and oblique hill-shading, and blue for water. In the new edition (1924) the number of colours is reduced to five—blue, green, black, and two shades of bistre.

The British Ordnance Survey, as we have said, was established in 1791, though the joint triangulation with France had been begun in 1784. On a scale of 1/63,360, the first sheet, in black and white with hachures to show relief, was published in 1801. In 1825 attention was directed to Ireland where, with a view to land valuation, a survey on a scale of six inches to one mile was begun and completed by 1840. This remained the original, from which smaller scale maps were reduced, till 1887, when a new survey was begun on a scale of twenty-five inches to one mile, and was completed in 1914. In the second half of the nineteenth century a new survey was embarked upon for Great Britain on a scale of six inches to one mile, which is now completed.

The first national map to be completed was that of Belgium, begun by Ferraris in 1770. A new survey was started in 1846 and completed in 1883, on a scale of 1/20,000, with a standard 1/40,000 reduced from it. In Switzerland, the cantons originally made separate surveys, but a federal commission in 1832 decided on a standard map on a scale of 1/100,000, the preparation of which was entrusted to General Dufour, after whom the map is named. This map, which is the most beautiful example of the hachure system of representing relief, is now superseded by a new map, on which relief is shown by contours, on a scale of 1/50,000 or 1/25,000.

In other European countries, the progress of national surveys was retarded by lack of political union or stability. The States of Germany each commenced separate topographic surveys in the nineteenth century, with no attempt to work on a uniform scheme. An Imperial Commission was formed in 1878, and decided on a general survey and the production of a uniform map with a scale of 1/100,000. Almost the whole of Germany has been mapped on this scale. In Italy, after political unification, the Istituto Geografico Militare was founded in 1875, and the survey of the whole country was completed by 1890 on an alternative scale of 1/25,000 or 1/50,000. In Austria-Hungary, though cartography had an early start (the first official establishment in Austria was formed after the Seven Years' War), a systematic large-scale survey was not embarked upon till 1869 (1/25,000) with a standard map of 1/75,000. This survey was completed in 1889.

In the United States the Geologic Survey, organized in 1879, is responsible for the topographic survey, and up to the present about 43 per cent. of the country has been covered on the scales of 1/31,680, 1/62,500, or 1/125,000. The British dominions are hampered by the vastness of their areas, though with the help of aerial survey rapid progress is being made in Canada.

Accurate topographic surveys are thus in existence to-day for nearly the whole of Europe (Spain is the chief exception with about one-third of its area mapped on a scale of 1/50,000), about half of the United States, eastern Canada, the whole of India and Burma (the trigonometrical survey begun in 1800), and the Japanese empire. Scattered areas with accurate surveys are situated around the large cities in the southern hemisphere, a great part of the Union of South Africa, and the countries bordering the Mediterranean Sea, North Africa, Egypt, Palestine, Syria, and Iraq. Less detailed surveys have been made for most of the surface of each of these countries, and the areas which are unmapped lie in the tundra wastes of Canada, central Asia, apart from the zone of the Trans-Siberian railway, China (with only very general maps on a scale of 1/4 million), the hot deserts and equatorial forests, and the Poles.

In the interests of systematic and comparable surveys international agreement is essential for the preparation of maps on a uniform system. Some forty years ago Professor Penck placed before the International Geographical Congress at Berlin, a scheme for an 'International Map of the World' on a scale of 1/1,000,000 (1/M.). The project was not taken up till 1909, when a special commission of delegates from the principal countries of the world met in London to draft a scheme as a basis for such a map. The scale, projection, general conventional signs, methods of representing relief, the classification of towns, style of printing, and the spelling of names were all decided upon, and the resolutions of this Committee still form the basis of the present map, though modified in detail by subsequent conferences in 1913 (Paris) and 1928 (London). Relief is shown by layer colouring, with contours, with the unit of vertical measurement in metres. The world map will consist of about 1,500 sheets excluding the seas, and of these about 10 per cent. have been published. These cover mainly most of Europe and south-western Asia, India, parts of South America, and Africa.

Modern atlas maps had their beginning in the establishment founded in Germany by Justus Perthes at Gotha in 1785, which published the first modern general atlas compiled by A. Stieler (1817–32), and the first physical atlas, prepared by H. Berghaus (1st edition, 1838–42). The latter when re-edited in 1887–92 included sections on geology and hydrography by Berghaus, meteorology by Hann, earth magnetism by Neumayer, plant distribution by O. Drude, and animal distribution by Marshall, and the distribution of man by Gerland. An English edition of the atlas was published by Keith Johnston with the aid of Petermann. The Stieler and, later, Andrée atlases held the field in the last century, but *The Times Atlas of the World*, prepared by J. G. Bartholomew, is now the finest general atlas of its kind. Since the publication of the later edition of the Berghaus physical atlas, others have been produced which will be noticed under the heading of the topics with which they deal (see section on development of climatology, Chapter XVI).

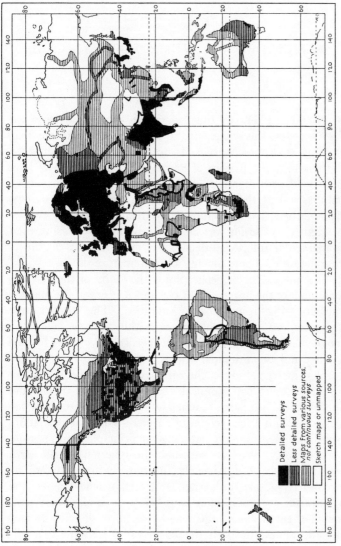

Detailed surveys

Less detailed surveys

Maps from various sources, not continuous surveys

Sketch maps or unmapped

FIG. 28. The Mapping of the World.

Chapter XV

ALEXANDER VON HUMBOLDT AND CARL RITTER (1800–60)

I

GEOGRAPHY, as the description of the earth, is the oldest, as the science of the interrelations between man and his environment, one of the youngest, of the sciences. Throughout the history of its development two views have been held with regard to its content, one of which has been emphasized at the expense of the other by some writers, while others have attempted to treat both aspects independently. Varenius was the only geographer to realize their fundamental interdependence, but even he was unable to complete his scheme owing to his premature death.

The one aspect of geography is the study of the earth as a member of the universe and as a unit; the other, the detailed encyclopaedic description of countries. The study of the universe and the earth's place within it was called by medievalists cosmography, of which geography proper formed a part, and the principal scientific advances in world geography, i.e. the study of the earth as a unit, were dependent on advances in astronomical and mathematical theory and speculation. But lack of factual knowledge did not deter the classicists and their scientific successors from speculation, based to some extent on mathematical principles and a passion for symmetry. We have seen how the sphericity of the earth was indirectly, but correctly, proved centuries before its circumnavigation; the existence of certain land-masses (the oecumene) was postulated long before their discovery, and the mathematical arrangement of parallels of latitude and climatic zones for both the northern and southern hemispheres were established long before navigators had disproved the popular conception of the inferno of the Torrid Zone, and opened the way for exploration beyond it. On the other hand, the description of countries from Strabo onwards, till after the time of Carl Ritter in the middle of the nineteenth century, remained encyclopaedic, with no definite aim, and with no principle of scientific description or of correlation and co-ordination of human and physical facts. The progress of this branch of geography was entirely dependent upon the production of accurate maps and the evolution of a definite concept of the aim, method, and scope of the geography of small areas.

The relations of cosmography and geography are clearly stated by Nathaniel Carpenter:

'Geographie is a science which teacheth the description of the whole earth. The nature of Geographie is well expressed in the name. For Geographie resolved according to the Greek etymologie signifieth as much as a description of the Earth; so that it differs from Cosmography as a part from the whole. For as much as Cosmography according to the name is a description of the whole world, comprehending under it as well Geographie as Astronomie. Howbeit, I confess, that amongst the ancient writers, Cosmographie has been taken for one and the self same science with Geographie as may appeare by sundry treatises merely Geographicall, yet entitled by the name of Cosmographie.'

The subject-matter of geography Varenius divided into general geography, dealing with the earth as a unit, and special geography, or the description of countries. But he considered that to such descriptions general world principles should be applied, the first statement to be found in the works of early geographers which foreshadows the scientific method of modern regional geography. To the more detailed study of areas Ptolemy applied the term chorography, and to the description of the very smallest areas, topography. These aspects of the subject were jettisoned by some writers as beyond its field, maintaining that chorography is distinct from geography. Varenius went half-way, by consenting to the inclusion of special geography, though he protested against, but agreed to include, as a concession to popular opinion, the description of manners, customs, government, &c.—a view with which modern geographers are entirely in sympathy.

Geography first claimed as its field the study of *all* terrestrial phenomena, but its newness lies in the narrowing of its field through the expansion of knowledge and the growth of new daughter sciences, which formerly lay within its sphere; and the systematization and correlation of the facts which lie within its more limited scope. These developments came partly as a result of the scientific progress after the middle of the eighteenth century in science and the humanities. Also knowledge of the character and distribution of physical and human phenomena throughout the world was greatly increased through the work of scientific explorers, of whom the greatest were naturalists: Alexander von Humboldt; Robert Brown, the botanist who travelled extensively in Australia and collected some 4,000 new species of plants; Charles Darwin, who travelled the world in the *Beagle* (1831–6); Thomas Huxley; Sir Joseph Hooker, who accompanied Ross to the Antarctic and travelled in northern India (1847–51); and Alfred Russel Wallace, who carried out epoch-making work on the flora and fauna of the Malay Archipelago. The improvement in cartographic methods, the initiation and, in many civilized countries, completion of national topographic surveys, the production of maps based on general surveys of the great hitherto unexplored continental interiors, improvement in astronomical observations and methods of land measurement, and the collection of data under the auspices of governments and learned societies—all these developments resulted in a vast accumulation of facts bearing on all aspects of science, which formed the raw material of the astounding scientific progress of what Wallace has called 'the wonderful century'.

Darwin's *Origin of Species* (1859) definitely established the doctrine of evolution, yet he was preceded by, and under constant obligation to, a number of scientists who had laid its foundations—Lamarck, who published his *On the Organization of Living Bodies* in 1801; Lyell, who laid the foundations of modern geology in 1830; and Malthus, who wrote an epoch-making work on *The Principle of Population* at the end of the eighteenth century. But during the first half of the nineteenth century, empiricism was the natural keynote to scientific progress for two main reasons. Firstly, this is the first stage in the systematization of new facts, in advance of the deduction of general laws. Secondly, the old idea of the origin of the earth, known as catastrophism, as opposed to the developing view of uniformitarianism, was still generally held. This was reflected in the predominance of the old cosmogony (as opposed to the evolutionary concepts) which, based upon the evident adaptation of all forms of life to environment, argued that divine design was responsible for such perfect order and harmony on the earth. This idea, which was expounded at length at the end of the seventeenth century by John Ray, an English zoologist, became well established in the eighteenth century, and reached its culmination in Butler's work on the *Analogy of Religion, Natural and Revealed, to the Course and Constitution of Nature* (1736) and in Paley's famous *Natural Theology*, published in 1802.

The theory of evolution preached by Kant and brought to fruition by Darwin and others, breathed new life into the scientific spirit, and in particular it resulted in the introduction of causal, as opposed to teleological interpretation into the science of geography, physical and human. This is the keynote of the development of geography in the second half of the nineteenth century, which witnessed the advent of the genetic interpretation of land forms, the birth of the science of geomorphology, and the study of the distribution and activities of man as *determined* by environmental conditions. The final, and present stage is characterized by the more critical study of human relations with physical environment, for it is now recognized that man is not a creature of his environment, but through conscious endeavour, according to his stage of development and social heritage, adjusts himself to it. The interpretation of this adjustment has its core in the small area with a distinct type of environment, to which the name 'region' is applied. Regional geography, or chorography—to use Ptolemy's term which was still current in geographical parlance at the end of last century and is still evident in the German term for the smallest geographical unit (*Chore*)—is the heart of modern geography, while the interpretation of physical and human characteristics within the regional framework is referred, as far as possible, to the principles of general world geography.

Alexander von Humboldt and Carl Ritter may be regarded as the founders of modern geography, the first for his researches and writings on the character and interrelationships of physical phenomena, and the second for his conception of the essential interdependence of man and nature. Both conceived geography as consisting of the description and systematization of terrestrial phenomena, and the influence exerted by these upon all forms of life on the earth's surface. Humboldt was in the main a physical geographer and his method was truly scientific, in that he tried not only to co-ordinate facts, but to establish laws to account for their character and distribution. Ritter, on the other hand, was primarily an historian, and his chief concern in geography was to show the dependence of the history of mankind on the physical environment. But the whole of his labours were vitiated by his philosophical background. He stressed man at the expense of environment and he made no attempt to establish causal relationships between the two. With no rigid method, owing to the lack of a concise objective, Ritter placed no limitation on the sphere of his subject. He rightly gives the keynote to the modern concept of geography, when he writes of geography that: 'It is to use the whole circle of sciences, to illustrate its own individuality, not to exhibit their peculiarities. It must make them all give a portion, not the whole, and yet must keep itself single and clear.' This definition, however, is too vague to permit a scientific treatment of a subject with ill-defined limits, and in view of his philosophical outlook Ritter, though laying the foundations of modern human and regional geography, and sowing the seeds of many modern geographical concepts, opened his work to criticism through his neglect of causal treatment of physical features, and his undue emphasis on the history of man.

II

Alexander von Humboldt was born at Berlin in 1769. With the intention of becoming a statesman he entered the University of Frankfurt-on-Oder in 1787, where he studied philosophy and law. Later he went to Göttingen, where he attended lectures by the great zoologist, Blumenbach. He disliked the prospect of his future vocation, and in 1791 his widowed mother permitted him to go to the School of Mines at Freiberg. Here he received tuition from the famous geologist, Werner, and he also made there what proved to be a lifelong friendship with the botanist and traveller von Buch. In 1792 he was appointed an officer of mines at Bayreuth, Ansbach, and Steben in the Fichtel Mountains. In 1793 he published his first work on the vegetation of the mines of Freiberg. The wide range of his early interests and the versatility of his powers are indicated by the publication in 1797 of *The Nervous and Muscular Irritation of Animal Fibre* after long experiments on the phenomena of muscular activity. During his early twenties Humboldt visited England (1790) and Vienna (1792 and 1797), and in 1795 he made a geological and botanical tour of Switzerland and Italy. A turning-point in his career came with the death of his mother in 1796, upon whom his education had devolved after his father's death in 1779. They had contemplated for him a political career. An innate taste for adventure, however, was enlivened by his acquaintance with George Forster, a member of Cook's expeditions, whom he met at Göttingen. After a short stay in Paris (where he devoted some time to the study of meteorology), Humboldt, with ample private means, decided to travel. In company with Aimé Bonpland, a botanist, he set out to Madrid, with the intention of joining a French expedition to Egypt, but instead sailed for South America. For the best part of five years (1799–1804) he travelled extensively in South and Central America, and carried out exhaustive scientific inquiries, the result of which on their subsequent publication laid the broad foundations of physical geography and meteorology. The first one and a half years were spent in the exploration of the Orinoco basin, during which time he discovered the connexion of the source of the Orinoco with a tributary of the Amazon. In 1802 he reached Quito, ascended Chimborazo to the greatest height yet attained, crossed the Andes several times, descended to Truxillo and followed the arid coast of Peru as far as Lima. From Cuzco he sailed for Guayaquil and Acapulco, and reached Mexico in 1803, where he carried out further extensive explorations. He revisited Cuba, and then made a hurried visit to the United States and returned to Paris in 1804. Here at the age of 36 he stayed for the better part of twenty years preparing the publication of his researches in South America. In 1827 he went to Berlin to the Court and became a favourite of Frederick William III and IV. In 1829 (May to November) at the age of 60, he made a hurried but extensive journey in the Russian empire from the Neva to the Altai. Between 1830 and 1848 he lived alternately in Paris and Berlin and paid occasional visits to England and Denmark.

Humboldt stimulated Murchison in his Russian geological researches in the Urals, which led to the classification of the Permian rocks, and he long pressed for the establishment of magnetic observatories all over the globe; this was finally partly achieved through the co-operation of the English and Russian Governments, and greatly added to our knowledge of terrestrial magnetism. His great life work, however, was his description of the physical world, published under the title of *Cosmos*, which he envisaged through the greater part of his life. He gave a series of lectures on this subject at Berlin in 1827–8 and these form 'the cartoon for the great fresco of the Cosmos', which covered the whole range of contemporary scientific knowledge. The first two volumes were published in 1845–7, the third and fourth in 1850–8, and the last, in 1862, after his death (1859).

Humboldt's South American researches covered twenty-nine volumes, with 1,425 maps and plates. It was published in six parts:

1. *Voyages aux Régions équinoxiales au Nouveau Continent.*
2. *Recueil d'Observations de Zoologie et d'Anatomie comparées,* faites dans l'Océan Atlantique dans l'Intérieur du Nouveau Continent et dans la Mer du Sud pendant les années 1799–1804 (published in co-operation with other authors).
3. *Essai politique sur le Royaume de la Nouvelle Espagne,* a description of the Geography of Mexico, its area, political divisions, physical geography, population, agriculture, manufacturing industries, commerce—with a discussion of the possibility of the construction of an isthmian canal.
4. *Recueil d'Observations astronomiques, d'Opérations trigonométriques et de Mesures barométriques* (1799–1804).
5. *Physique générale et Géologie,* the geography of plants and physical geography based on observations and measurements between Lat. 10°, N. and S.
6. *Plantes équinoxiales:* collected in Mexico, Cuba, Caracas, Cumana, Barcelona, the Andes of New Granada, Quito, Rio Negro, Orinoco, and the Amazon.

As a result of his Siberian travels in 1829 Humboldt published a book on *Fragments de géologie et de climatologie asiatique* (2 vols., 1831), followed in 1843 by a more comprehensive work on *Recherches sur les chaînes de Montagnes et la Climatologie comparée* in which he deals with the geology, structure, astronomical records, and climate, and gives accounts of the mines of the Urals, and the gold-fields of Siberia.

That Humboldt was not merely a physical geographer is shown by the wide range of topics covered in his South American labours, while in 1807 he published a work in collaboration with Bonpland on the geography of plants, and later he established himself in the forefront of critical historians by the publication of *Examen critique de l'Histoire de la Géographie du Nouveau Continent et des Progrès de l'Astronomie nautique dans les XVe et XVIe siècles*. It was while engaged in this research that Humboldt, in Paris, discovered the famous map of the New World prepared by Juan de la Cosa.

The aim of Humboldt's *Cosmos*, 'a work', he says in his preface, 'whose undefined image has floated before my mind for almost half a century' is best summed up in his own words: 'The principal object by which I was directed was the earnest endeavour to comprehend all the phenomena of physical objects in their general connexions and to represent nature as one great whole, moved and animated by internal forces.' The foundations of physical geography are laid in the section on terrestrial phenomena, and its arrangement is remarkable for its similarity to the General Geography of Varenius, whose work Humboldt highly praises in the introduction.

The general features of the atmosphere are discussed under the heads of variations of atmospheric pressure, the climatic distribution of heat, the distribution of humidity, and the electric condition of the atmosphere. Among the more important contributions of Humboldt to physical geography mention should be made of isotherms, which, for Humboldt, were lines joining places with the same average annual temperature; lines joining places with equal winter temperatures he called isochimenals and summer temperatures isotherals. Isotherms, in Humboldt's words, 'serve as one of the main foundations of comparative climatology'; and an extension of the same principle permits numerous other quantitative phenomena to be shown in the same way—as was attempted for the first time in Berghaus's Atlas (1838–42). From the world distribution of temperature thus shown Humboldt noted the contrasts between the east and west coasts of the continents between lats. 40° and 60° and related them to the direction of the prevailing winds. He noted the difference between insular and continental climates, and he discusses generally the reasons why the isotherms do not run parallel to the parallels of latitude. He paid much attention to the vertical as well as the horizontal distribution of temperature.

'The law of the decrease of heat with the increase of elevation at different latitudes is one of the most important subjects involved in the study of meteorological processes, of the geography of plants, and of the various hypotheses that relate to the determination of the height of the atmosphere. In the many mountain journeys which I have undertaken, both within and without the tropics, the investigation of this law has always formed a special object of my researches.'

A whole volume is given to the investigation of volcanoes, or 'Fire-Emitting Mountains', extinct and active. They are arranged into linear clusters and their description correlated with subterranean fissures, and an attempt is also made to show that the subterranean centres of volcanic activity are in communication with each other. Humboldt also shows how volcanic action produces a distinct type of rock (for the general view was that all rocks were of aqueous origin) and how such action metamorphoses those already in existence.

There is also discussion of the distribution of islands and archipelagoes, ocean depths, the dependence of currents on prevailing winds, and the modification of density of sea water in consequence of differences in temperature and relative quantity of saline contents at different latitudes and depths. The east-west equatorial current is noted, to which Humboldt correctly assigns the origin of the Gulf Stream.

The distribution of plants and animals is also dealt with in *Cosmos*, and more fully in *Ansichten der Natur* (*Aspects of Nature*, trans. by Sabine). He deals with the horizontal and vertical distribution of organic forms, the distribution of plants being carefully related, as far as observations at that time would allow, to the distribution of temperature; and it is noted that certain isotherms correspond with the limits of certain plants and animals.

Volume ii of *Cosmos*, entitled *Incitements to the Study of Nature*, is a subjective treatment of the development of knowledge of the physical world, while volume iii deals with the History of Science. In the first volume Humboldt puts forward his general arguments and main conclusions; the detailed evidence on which these are based comprises the last three volumes.

Humboldt's *Cosmos* is a monument of compilation, and in it are embodied the conclusions drawn from a lifetime of travel and diligent research. His method, in his words, consisted in 'the art of collecting and arranging a mass of isolated facts and rising thence by a process of induction to general ideas'. Humboldt was unable to incorporate new developments in science owing to the long period which elapsed between the publication of the first and last volumes. For instance, Helmholtz's Conservation of Force, Joule's Mechanical Theory of Heat, and the spectrum analysis developed in the 'sixties were ignored, or partially embodied or criticized in the work, and in certain respects blighted some of his theories. Yet, in the words of Bruhnes, 'The book stands out unquestionably as the most comprehensive compendium of modern science, and as the most complete history of its development that has ever been attempted'. In the sphere of geography Humboldt's principal contributions were his investigations of horizontal temperature distribution and the method of its representation by isotherms, the vertical decrease of temperature in different latitudes, his correlation of plant distribution with physical conditions, investigations on the distribution of terrestrial magnetism, and the distribution and causes of volcanic activity. Finally, in his descriptions of Mexico and the llanos of the Orinoco basin, he compiled the first systematic geographical studies of separate regions.

Humboldt laid the foundations of systematic regional descriptions, and the general comparative study of like forms and regions on the earth's surface. He developed the principle of causation in geographical interpretation, realizing the essential interdependence of natural phenomena and their influence on man. Thus he writes in *Cosmos*: 'Whatever causes diversity of form or feature on the surface of our planet—mountains, great lakes, grassy steppes, and even deserts surrounded by a coast-like margin of forest—impresses some peculiar mark or character on the social state of its inhabitants. Continuous ridges of lofty mountains with snow impede intercourse and traffic; but where lowlands are interspersed with discontinuous chains and with groups of more moderate elevation, such as are happily presented by the south-west of Europe, meteorological processes and vegetable products are multiplied and varied; and different kinds of cultivation, even under the same latitude, give rise to different wants, which stimulate both the industry and the intercourse of its inhabitants.' The spirit of his comparative method is excellently illustrated by his essay on steppes and deserts in *Aspects of Nature* (vol. i) in which he remarks: 'It is a highly interesting though difficult task of general geography to compare the natural conditions of distant regions, and to represent by a few traits the results of this comparison' —after which he deals with the form and latitudinal extent and winds of the continents of Africa and South America to account for the contrasts in their covering of vegetation. Humboldt constantly sought correlation of facts—laws and consequences. As de Martonne has written: 'Whatever phenomena he studied, relief, temperature, vegetation, Humboldt did not merely treat each individually as a geologist, meteorologist, or botanist. His philosophical outlook carried him farther. It led him at once to the observation of other phenomena; he sought causes and distant consequences, even including political and historical facts. Nobody has shown with more precision how man depends on the soil, climate, vegetation, how vegetation is a function of physical phenomena, and how they all depend on each other' (*Traité de Géographie Physique*).

III

Carl Ritter had a career markedly contrasted to that of Humboldt. He was a teacher and a scholar, and travelled little. His training was mainly in the humanities. He drew much of his inspiration in physical geography from Humboldt and, adding his own philosophical concepts and bias towards history in the study of human activities and development, he laid the foundations of the new science.

Ritter was born in 1779 near Magdeburg. When he was six years of age he went to a small school where the teaching was based on Rousseau's principles, at Schnepfenthal near Gotha, and here he stayed till he was 17. Then he went to the University at Halle to train for a teacher, during which time he was supported by a wealthy banker of Frankfurt, Hollweg by name. While at the University he studied botany and mineralogy, but his main interests were in mathematics and philosophy; later he turned to history, pedagogy, physics, and chemistry. In 1798 he became private tutor to the two sons of the Hollweg family at Frankfurt and, while occupying this post, his studies drifted to the classics. On several occasions he visited Switzerland and met and was inspired by Pestalozzi, but the turning-point of his career seems to have been his contact with Humboldt, whom he met in 1806. While at Frankfurt Ritter became increasingly absorbed in a book he was writing on physical geography. In this work he says that 'many riddles have been solved. . . . I have gone farther than my predecessors, particularly in the study of the marine currents, the course of the winds, the division of the continents into mountain lands and plains, the formation of river valleys, the laws of climate, and the diffusion of minerals.' In 1804 he published his first treatise on Europe, in the preface of which he writes: 'The earth and its inhabitants stand in the closest mutual relations and one element cannot be seen in all its phases without the others. On this account history and geography must always go hand in hand. The country works upon the people and the people upon the country.' In connexion with this book Ritter also prepared maps of Europe, showing its mountain systems and the distribution of vegetation and cultivated plants in relation to climate, animals, and languages.

In charge of his two young pupils Ritter visited Switzerland and Italy, and later he stayed at Göttingen for two years, during the university career of his elder pupil. Here he attended courses in jurisprudence, medicine, mineralogy, geology, and botany. The first volume of his masterpiece, the *Erdkunde*, was published in 1817. His reputation was made, and he soon received a post at the gymnasium at Frankfurt. After a short time he moved to Berlin, where, at the invitation of the State, he gave instruction in 'military statistics'. In Berlin Ritter was in his prime. He delivered regular public lectures which attracted vast numbers of young students, and through a quiet but strong and inspiring personality he exercised a tremendous influence on his hearers in public, and on those with whom he had personal contacts. The best of his courses, which has been translated into English, was on *General Comparative Geography*, but in addition he lectured on Europe, Asia, and the classical countries. For some ten years Ritter devoted all his energies to his lecturing, but in 1831 he returned to his writing, and in 1832 appeared the second volume of the *Erdkunde* on Asia. Between 1832 and 1838 he published a further six volumes, eleven between 1838 and 1859, and lost in a welter of detail, through vagueness of conception, the work was far from completed when he died in 1859.

With regard to the content of geography he writes: 'The very word geography, meaning a description of the earth, has unfortunately been at fault, and has misled the world; to us, it merely hints at the elements, the factors of what is the true science of geography. That science aims at nothing less than to embrace the most complete and the most cosmical view of the earth; to sum up and organize into a beautiful unity *all that we know of the globe*. . . . Geography is the department of science that deals with the globe in all its features, phenomena, and relations, as an independent unit, and shows the connexion of this unified whole with man and with man's Creator' (*Comparative Geography*). The central principle of geography, he claims, is 'the relation of all the phenomena and forms of nature to the human race'. Thus, as already quoted, geography must draw upon all the kindred sciences. 'It is to use the whole circle of sciences to illustrate its own individuality, not to exhibit their peculiarities. It must make them all give a portion, not the whole, and yet must keep itself single and clear.'

The *Erdkunde*, Ritter claimed, would definitely establish geography as a science. The great superiority of his book over those of his predecessors was, he says, because 'my aim has not been merely to collect and arrange a larger mass of materials than any predecessor, but to notice *the general laws* which underlie all the diversity of nature, to show their connexion with every fact taken singly, and to indicate in a purely historical field the perfect unity and harmony which exist in the apparent diversity and caprice which prevail on the globe, and which seem most marked in the mutual relations of nature and man. Out of this course of study there springs the science of physical geography, in which are to be traced all the laws and conditions under whose influence the great diversity in things, nations, and individuals first springs into existence, and undergoes all its subsequent modifications'. (Letter quoted by Gage, pp. 143–4.)

The character of the great work, however, though bearing the seeds of method, suffered from the vague philosophical outlook of its writer, which could not be applied in detailed geographical treatment. His conception was philosophical, and his method descriptive, with an attempt at systematization, but not causation.

The outstanding feature of Ritter's conception of the scope and aim of geography, which influences the character of all his work and proved to be the principal deterrent after his death to the advancement of the science, was his teleological conception of the mutual relations of man and nature, a philosophy which, as noted above, was merely a reflection of the prevailing thought of his age. Ritter believed the earth to be an organism, made even in its smallest details with divine intent, to fit the needs of man to perfection. In his view, 'as the body is made for the soul, so is the physical globe made for mankind'. The central theme underlying all his work is that the earth is not a mere dwelling-place for nations, but the very material out of which life is woven. 'The earth is the garment in which the soul clothes itself, the body wherein the spirit formed by God, must move' (Gage). The idea is again expressed in such phrases as follows, taken from his *Comparative Geography*: 'The constitution of the globe is incontestably coincident with a plan to perfect and preserve man'; 'The distribution and shape of the continents and their relation to each other, show the workings of the Divine mind in the interests of humanity'; and again, 'The individuality of the Earth must be the watch-word of recreated geography', this individual life of the earth being of divine origin. Arnold Guyot, Ritter's greatest disciple, who carried his teaching to the other side of the Atlantic, expresses the ultimate goal and primary objective of geography as follows: '. . . inorganic nature is made for organized nature and the whole globe for man, as both are made for God, the origin and end of all things'; and 'it is thus that science takes in the whole of created things, as a vast harmony, of which all the parts are closely connected together and presuppose each other' (*Earth and Man*, 1873). The goal of the science is expressed briefly by the same author as follows:

1. That the forms, the arrangement and the distribution of the terrestrial masses on the surface of the globe, accidental in appearance, yet reveal a plan which we are partially unable to understand by the evolution of history.
2. That the continents are made for human societies, as the body is made for the soul.
3. That each of the northern or historical continents is peculiarly adapted by its nature to perform a special part which corresponds to the wants of humanity in one of the great phases of its history.

By a complete understanding of the mutual relations of man and environment, Ritter envisages the time when it will be possible for men 'sending their glance backwards and forwards, to determine from the whole of a nation's surroundings, what the course of its development is to be, and to indicate in advance of history, what ways it must take to retain the welfare which Providence has appointed for every nation whose direction is right and whose conformity to law is constant'.

Ritter's physical geography was descriptive and not causal; he did not even make full use of the available knowledge of his time. In the form of the lands he merely saw the stage for the human drama, a stage set to perfection for its performance. Out of the confusion of earth features he read nothing but order, a premeditated divine arrangement and adaptation to one end—human progress. And it was the latter which formed the great bulk of his researches. But Ritter was fully alive to the changing significance of natural features to man, according to the stage of human development. Historical geography did not mean to him merely 'a certain amount of historical facts connected with a geographical spot, but the variety of functions performed by the same geographical elements, or the same natural regions, in the different ages of civilization' (Gage). He realized that the potency of such functions varies, since 'they depend on the power of the cultivated nations to make use of these elements furnished by nature, as instruments for the particular work which these nations are called upon to perform in history'.

His conception of the influence of the diversity of physical features upon the course of human development is stated by Gage as follows: 'Climates are diversified, the formation of distinct nations favoured, a greater variety of human faculties called into action, mutual relations and reciprocal influences increased, which in the indented continents unfold the hidden powers of man to a degree unknown in the continents less favoured in this respect.' Thus the progress of civilization in Europe he ascribed to its position as a peninsular prolongation of Asia, its proximity to the site of the early riverine civilizations of south-western Asia, and its configuration, the interdigitation of land and water. The development of European civilization he divided into the three stages, Riverine,

Mediterranean, and Atlantic, showing how the relative value of geographical elements has changed with human progress.

Ritter also appreciated the influence exerted by man on nature, his 'subjugation of nature', as instanced by the development of civilization in Egypt based upon irrigation from the Nile, the conquest of natural barriers such as the high passes of the Alps, and of the oceans and the great rivers by the development of navigation. In fact, 'the changes which Art is yet to effect on our Globe are beyond all possible computation and it may be said beyond all possible exaggeration'.

In the sphere of physical geography, Ritter, though, as in most lines of research, he did not originate the idea, first developed the comparative method in geography. This method simply consists of the comparative study of simple and similar earth features, with a view to establishing some kind of order out of the apparent chaotic distribution of land and water. Ritter, however, only adopted the method in his comparison of the form and configuration of the continents considered as units, with a view to comparing and accounting for the progress of civilization. His classification of the major features of the lands is purely descriptive. He divided land forms into four main groups, which he conceived to be systematically grouped together to give a similarity of general build to each of the continents. First are the highlands and plateaux, which are of two orders, the first with an average elevation of 4,000 to 5,000 feet, and the second of lower average elevation. Second are the mountains, which are divided into five groups:

(a) parallel mountain chains, e.g. the Jura and Himalaya;
(b) diverging or converging mountain chains, e.g. the Eastern Alps and Northern Rockies (divergence), and the mountain knots of the Andes (convergence);

(c) ranges radiating from a central nucleus, e.g. Auvergne, south-west Alps.
(d) ring-shaped systems, e.g. Transylvania and Bohemia;
(e) Cross mountains, in which two or more ranges meet at high angles, e.g. the Hindu Kush and Himalayas, the Kwen Lun and the Pamirs.

Third are the lowlands, with an elevation below 400 feet, an example of which is the great European plain, which is considered to be an uplifted sea floor. Fourth are the regions of transition between highlands and lowlands, called lands of gradation or terrace lands.

This classification, summed up in his *Comparative Geography*, forms the basis of his treatment of the continents. Superficial as the scheme may be, it afforded a new *regional* method of description, as distinct from the usual modern method of dealing by political units. The first continents Ritter studied were Asia and Africa, and it was on the basis of their configuration that he arrived at his classification of land forms, and conceived each continent as being built on a similar plan. At the heart of each continent is a highland or tableland around which are grouped subsidiary physical units. First are the terraces or lands of gradation, and they are followed by peripheral lowlands. Finally, beyond and surrounding the whole structure 'as with a garland of brilliant flowers', are the peninsulas and islands.

On the basis of this classification it is important to note Ritter's system of regional divisions. For this purpose Africa has been selected. He describes each division, proceeding from the colder and less favoured to the warmer and richer regions, in the order of physical conditions, people, and historical development and present conditions. He divides the continent into four main units:

1. High Africa, or the Tableland.
2. The terrace lands, transitional between 1 and the low-

lands and containing the greater part of the river valleys.

3. The isolated plateaux of Atlas and Barca.

4. The lowlands of northern Africa (Sahara).

The subdivisions are as follows:

I. (a) Southern margin and its terraces, descending to the Cape of Good Hope.
    (i) High tableland or Orange River.
    (ii) Middle terrace of the Karroos.
    (iii) Lower terrace or Shore Region.

  (b) Eastern border of the highland with terraces.
    (i) Kaffir coast.
    (ii) Coast of Sofala and Mozambique.

  (c) Northern Margin.
    (i) High terrace of Kaffa and Narea.
    (ii) Tableland of Abyssinia with surrounding terraces.
    (iii) Lowland of northern Africa.

  (d) Western Margin.
    (i) South-west coast (Cape Negro to Gonzales).
    (ii) Zaire River basin.
    (iii) Headland of Ambos.
    (iv) Upper Sudan.

  (e) West Africa.
    (i) Tableland of Mandingoes.
    (ii) Rivers Senegal, Gambia, Niger, and the Kong Mountains.

II. The great river systems and their regions.
    (Terrace lands.)
    Orange River.
    Terraces and streams of Middle Africa. (Senegal, Niger, Nile.)

III. Atlas and Barca tableland.

IV. (a) Eastern Sahara (with oases).
    (b) Western Sahara and Sahel.

Ritter's conception and philosophy of geography was vague, and his method of detailed treatment, owing to this lack of precision of aim, together, of course, with the lack of scientific data especially for the treatment of the dark continent, resulted in the inclusion of masses of detail irrelevant to geography proper. Stressing more the history of man than the theatre of his evolution, he neglected physical and mathematical geography which had no place in the published part of his *Erdkunde*. The fact that he eventually hoped to publish such a volume towards the end of the work is again indicative of incorrect procedure, for sound general principles should come before regional description. He was concerned less with the causes of natural conditions than their effect upon man, and even these were not given clearness of exposition, analysed on a scientific basis of cause and effect.

It is easy to criticize Ritter in the light of modern thought, but in his work lay the seeds of many geographical concepts. He clearly envisaged the nature of the science, though not its scope. He recognized the mutual relations of man and his environment. He realized the need for divisions other than political for detailed treatment. He suggested the study of maps for the comparison of surface features, and though he only applied this method to the general outline and build of the continents, it was greatly enlarged upon by his successors, and is fundamental to modern geography. Ritter had the true geographical outlook, though the content of the subject he defined has since been gradually narrowed down. His two great handicaps were, first, his teleological philosophy, which led to a dead end and prevented rational treatment; secondly, the lack of appropriate data for detailed work. He visualized a scheme but failed to carry it out, through his historical and teleological bias. Without this bias, with a true scientific spirit, and by his method of writing—relegating all detail from his general essays whereby the main argument was always clear to view—Humboldt served modern geography where Ritter failed, through his systematic method of regional description, comparative study of terrestrial phenomena and exposition of the interdependence of them all. These ideas Humboldt concisely and scientifically demonstrated; Ritter visualized them but failed to apply them.

Chapter XVI

THE DEVELOPMENT OF PHYSICAL GEOGRAPHY

THE keystones of modern scientific thought are Charles Lyell's *Principles of Geology*, which definitely established the theory of the gradual development of the earth's surface features, or the doctrine of Uniformitarianism, and Charles Darwin's *Origin of Species* published in 1859, the year of the death of Humboldt and Ritter. Darwin brought forward a vast body of evidence to prove that organic life in all its forms had evolved slowly, and not by a series of cataclysms, each of which killed off one phase of life and witnessed the establishment of another distinct in type and advanced in form. In their respective inorganic and organic spheres these two giants of science established the doctrine of evolution, and proved that the development of the earth's surface and its living organisms could be interpreted in the light of observable facts, in the existing processes of nature, inorganic and organic. The catastrophic theory of the origin of the earth's surface features and all its life forms, and the associated teleological belief of their mutual adaptation through divine intent, now gave place to the more rational evolutionary concept. The existing features of the earth came gradually to be regarded as passing phases in a process of constant growth and decay. The life of the globe was formerly conceived to be naturally fitted to the environment (or in the case of animal and plant life, endowed with distinctive and apparently useless forms, merely for the delectation of man), but the new philosophy held that all organic life was moulded by the environment, and all organic traits were but responses (either past, in the case of the survival of useless traits, or present) to the necessity for existence.

Geography, which, in its early stages in this century,

ambitiously claimed as its field the distribution of all phenomena and their relation to the physical features of the earth's surface, has in its nineteenth-century development, a dual aspect: firstly, the systematization of the earth's features and their genetic interpretation; secondly, the relation of all forms of life, biological and human, on the earth to its natural features. The great difference between these two aspects is that earth forms are the product of internal forces or earth movements, and the moulding action of the agents of erosion. Man (and to a lesser extent animals and plants) is an active agent and reacts upon his environment, and though the potency of human control of physical conditions was early realized in a general way, the inevitable initial trend was that, in the desire to apply the same genetic interpretation to human life and development, man should be envisaged as the creature of his environment, to the exclusion of the influence of all other factors.

Physical geography is concerned with the description and development of the earth's surface features. With the accumulation of data by scientific travellers, and the progress of accurate cartography, the first stage in establishing some kind of order into the multitudinous variety and apparent chaos of surface features was through the comparison of features, large and small, over the earth, and an attempt to group them according to similarity of form—that is, the method of treatment was essentially empirical and descriptive. Genetic interpretation, or the study of the evolution of land forms, and systematic classification on this basis, were necessarily dependent upon the advance of knowledge regarding the nature of earth movements and the agents of erosion, particularly water and ice. While the former showed a progressive development during the nineteenth century, the efficacy of the work of running water and the nature of the cycle of erosion were not fully worked out till the last third of the century, and even the existence and effects of a Quaternary ice sheet over northern Europe and America were disputed till the 'seventies. The genetic interpretation and classification of land forms is so intimately associated with progress in these two spheres that the main stages in their development will be summarized before dealing with the development of physical geography proper.

I

Among those who speculated on the causes of the configuration of the earth before the nineteenth century, mention should be made of Descartes, Leibnitz, and Buffon, who ascribed the origin of ocean basins, continents, and mountains to the wrinkling and fracturing of the earth's solid crust, and to the withdrawal of the surface waters into the cavities of the earth's surface; while others held that the continents and mountain-systems owed their origin to volcanic forces.

At the end of the eighteenth century, Pallas and De Saussure attempted to explain the origin of mountains by investigating their stratigraphy and structure. Pallas (1741– 1811) published in 1777 his *Consideration of the Structure of Mountain Chains*. He divided mountains according to their character and age into three groups, according to the time of their upheaval from the primeval ocean. Those mountains with granite cores are the oldest and were never submerged by the ocean. Schist mountains were formed before the advent of living organisms, as they contain no fossil life. Limestone and other ranges of fossiliferous rocks are the youngest, and were due to the last volcanic eruptions. At the time of these upheavals great cavities were formed and occupied by the oceans, and the latter sometimes flooded the continents. De Saussure (1740–99), who gave his life to the study of the French-Swiss Alps, had views similar to those of Pallas, and he also added to the progress of physical geography by his measurement of heights of snow lines, determinations of the rise of temperature in the ground and of the depth of lakes, and investigations of glaciers and of the altitudinal distribution of plants.

The development of modern thought on earth movement begins with the Scottish geologist James Hutton (1726–97) and his principal disciple, John Playfair. He maintained that the configuration of the earth was due to the expansion of molten rock in its interior, and that volcanoes provided an outlet, like safety valves, for the molten rock, and thereby prevented excessive expansion and upheaval. The idea was accepted and elaborated by Leopold von Buch, and later by Lyell.

But the most enthusiastic exponent of the Vulcanist doctrine was Elie de Beaumont. In 1852, in a three-volume work, *On Mountain Systems*, he explains mountain-folding by the contraction of the earth's interior and the consequent wrinkling of the crust, owing to its surplus area for a diminished volume. The wrinkling resulted in either excrescences or folding, due to lateral compression. Sometimes the deep-seated molten rock would burst through the upheaved crust, a fact which accounts for the occurrence of granite in the heart of many mountain-chains, while their flanks consist of sedimentary rocks. De Beaumont determined the age of mountain-folds from a comparison of the contorted strata of the mountains, and the horizontal strata of the younger rocks on their flanks, and each mountain-system, he argued, corresponded with one of the great cataclysms postulated by Cuvier in his idea of the development of life. At a later stage, wandering from the field of induction to speculation, he attempted to correlate the age of mountains with their strike, and, from a study of the mountain-systems of Europe, he formed a general geometrical law of orientation for the mountains of the earth.

Professor James Dwight Dana (1813–95), an American geologist, first clearly enunciated the theory of horizontal compression, due to the earth's contraction, to explain the origin of mountains. He assumed that the centripetal movement of the earth's crust resulted in horizontal pressure which folds the crust, on the margins of the continents, into ridges and corresponding troughs. The former he called ge-anticlinals, the latter geo-synclinals. The geo-synclinals lie on the edges of the continents and are therefore the centres of great deposition, and from them the mountain-systems are upheaved and incorporated with the continents. Vulcanicity is most active where the crust is weak. With the cooling of the earth and the increased thickness of its crust, the areas of oldest disturbance would be most stable, and the areas of recent tectonic disturbance least stable. Hence, the concentration of the chief volcanic areas along the line of recent folded mountains—especially along fault lines.

In 1875 Lowthian Green (*Vestiges of the Molten Globe*) put forward his Tetrahedral Theory of the distribution of continents and oceans, on the basis that a contracting sphere tends to assume the form of a tetrahedron, a body enclosed by four equal equilateral triangles. A sphere has a minimum, and a tetrahedron a maximum surface area for a given volume; therefore, a slight contraction of the earth's interior would result in a tendency for the wrinkled crust to assume the form of a tetrahedron, the faces of its sides containing the seas, and its edges the land masses.

Professor Edward Suess in 1875 published a remarkable work on the origin of the Alps, in which he discusses mountain-folding in general. He summarizes his conclusions as follows: the strikes of mountains do not always run parallel with the greater circles of the earth (as suggested by Elie de Beaumont), but may be diverted by various obstacles (resistant crustal blocks, old mountain masses, e.g. Bohemia and the south Russian block); and he agreed that the great fold-mountains have their origin in the geo-synclinals as suggested by Dana. Dana's theory of lateral compression, based upon study of the Appalachians, Suess showed was also applicable to the formation of the mountains of Europe. These ideas Suess embodied in his great work entitled *Das Antlitz der Erde* (*The Face of the Earth*) (1883–1901) in which the chief aim is to explain the actual configuration of the earth's surface in terms of past changes in oceans and continents. He held that the plan of the earth is due to two forms of crustal movement:

(1) subsidence over wide areas giving rise to the oceans,
(2) folding along narrow belts giving mountain-ranges on the continents.

This theory is opposed to that put forward by C. Lapworth in 1892. This geologist ascribes the harmony of surface configuration to two series of crustal waves running from east to west, and north to south, thus giving rise to six continents arrayed in three groups separated in the centres by minor depressions, and from each other by major depressions.

While some of Suess's views have not withstood criticism in the light of recent research (Argand, Staub, Hobbs, Wegener, Joly, and others) the bulk of this brilliant work, based upon personal investigation and the arduous lifelong task of collecting the writings of a century's workers, is still the best standard reference for the study of the physical evolution of the earth and its parts.[1]

II

The nature and potency of the two chief agents of erosion, running water and glaciers, were not fully worked out on a scientific basis till the second half of the century.

In the eighteenth century Buffon attributed the excavation of river valleys to the action of submarine currents during the retreat of the oceans, after the flooding of the lands. The diluvial theory, that the Mosaic flood was the last of a series,

[1] For recent thought on the 'History of the Earth's Surface', see an editorial article in *The Geographical Teacher*, Autumn, 1925.

and was responsible for the shaping of the existing surface features, was generally held, in various modified forms, by most of the geologists of the early nineteenth century—De Saussure, Pallas, Werner, and Elie de Beaumont. Buckland even coined the term 'denudation' for the scouring of the continents due to this universal flood.

But Hutton and Playfair started the investigation of the problem on scientific lines—though in some respects they had been anticipated by the Italian geologists, Guettard and Targioni. In 1785 Hutton presented his paper *The Theory of the Earth, or an Investigation of the Laws Observable in the Composition, Dissolution and Restoration of Land upon the Globe*, in which he laid the foundations of the uniformitarian doctrine, by putting forward a theory for the origin of the surface features of the earth, and explaining their characters as due to the slow process of decay by the action of running water and atmospheric agencies.

It was Sir Charles Lyell, however, who finally laid low the catastrophic theory by the publication of his *Principles of Geology*, 'an enquiry how far the former changes of the earth's surface are referable to causes now in operation'. This work consists of four volumes, on climatic variations in the history of the earth, the agents of denudation, coral reefs, and historical geology. In the second volume he ascribes valleys to river erosion, after which this explanation came to be currently accepted.

But it was generally believed by the uniformitarians that rivers only roughened the surface by carving valleys; they had no idea of progressive erosion resulting in the ultimate disappearance of a former mountain-system, till only its stumps remained to form a rolling surface, almost a plain (peneplain). Such abrasion was considered to be only possible by sea action, and plains of marine denudation were early recognized by Ramsay (1847) and later by Von Richthofen (1882).

The first stage in the study of river action was naturally the study of individual rivers, and the principal contributions were made by pioneers in the Far West of the United States, where the grandeur and diagrammatic simplicity of land forms facilitated the relatively easy interpretation of their dependence on structure and the action of running water upon them.

In the middle of the century, J. L. Lesley showed the dependency of the topography of Pennsylvania on structure in his *Manual of Coal and its Topography* (1856). He did not, however, appreciate the significance of river erosion, and he did not present the land forms as members of an evolving series. This idea was to be worked out by later workers in the same field and in the Far West, west of the 100th meridian.

Gilbert, an American geologist, in his *Geology of the Henry Mountains* (1877), established the chief laws of river action in the erosion of valleys. His conclusions were corroborated by the researches of Powell and Dutton in the Grand Canyons of the Colorado, though fracturing was still believed by some till the 'eighties to be the cause of these phenomenal valley gorges. These geologists demonstrated the process of backward erosion during the excavation of a valley, and that there is a definite relation between the gradient and erosive capacity of a river. W. M. Davis, from his studies of river development in the Pennsylvanian section of the Appalachian Highlands, and on the Atlantic coastal plain (see his *Geographical Essays*) elaborated the stages in the development of a river valley: the young stage, when the river has a narrow deep channel with rapid gradient and waterfalls; the mature stage, with a broader valley and a decreased angle of declivity, and interlocking spurs; the senile stage, when it has reached its base level. Crustal movement in any part of the valley results in its 'rejuvenation', and the cycle begins again. Thus was elaborated in the 'eighties the theory of the base level and the cycle of erosion.

Gilbert first used the terms consequent and subsequent to rivers, on the basis of river development on a dome, the former flowing from the summit of the dome across the strata, and the latter, appearing with the erosion and exposure of the strata, running parallel to the strike. The term 'obsequent' was later suggested by Jukes from a study of the valleys of south-western Ireland (1862). In 1865, Medlicott suggested that the rivers Indus and Brahmaputra had existed before the upheaval of the Himalayas and that, owing to the slowness of the folding, they had been able to maintain their courses. The same idea was expressed by J. W. Powell in explaining the course of the Green River through the Uintah Mountains, and the Colorado through the Arizona plateaux, and he uses for the first time the terms 'antecedent' and 'superimposed', as applicable to such rivers.

The significance of ice as a geological agent was realized much later than that of running water, owing largely to its limited distribution and inaccessibility. Investigations of the geological action of the Swiss glaciers, and the deduction therefrom of a former extensive glaciation of northern Europe and North America, was one of the most brilliant discoveries in the field of geology in the nineteenth century.

After prolonged study in the Alps Louis Agassiz in 1837 expressed the view that before the upheaval of the Alps their site had been extensively covered with ice which, on its disappearance, had left traits in the form of erratic blocks and polished rocks; he also described the principal characteristics of a typical valley glacier. Shortly afterwards (1840) he enunciated his theory of an Ice Age, based upon the existence of similar phenomena in Scotland and Scandinavia. This ice sheet, he suggested, extended over Europe and the Mediterranean as far as the Atlas Mountains, northern Asia, and North America. In his *Études sur les Glaciers* (1840) Agassiz modified these views, for he now believed that the glaciation of the Alps was distinct from that of northern Europe, and that the glaciation took place subsequent to the upheaval of the Alps. In 1841 Charpentier in his *Essai sur les Glaciers* gave a brilliant description of glacial phenomena, and conclusively proved that erratic blocks were due to glacial action.

Meanwhile, under the leadership of Leopold von Buch, a German, geologists in general argued that the erratic blocks of the northern European plain had been carried and laid down by vast floods coming from Scandinavia. This view was also accepted by a group of British geologists (including Darwin) who, in the early 'forties, enunciated the 'drift theory', by which the existence of erratics and boulder clay was ascribed to floating icebergs drifting from the north polar regions; the term drift then came to be applied to these deposits in this country. For a time this theory completely eclipsed Agassiz's idea of an extensive European ice sheet. It received its death blow in the 'seventies, however, from the researches of Sir Andrew Ramsay, who proved that most of Great Britain had formerly been submerged beneath an ice sheet, and from similar conclusions drawn from researches in Scandinavia.

Since then German geologists have been to the fore in the study of glaciation, past and present, and the works of Penck and Bruckner should be particularly noted. In their classic work on *Die Alpen im Eiszeitalter* published in three volumes (1901–9) they showed how climatic fluctuations characterized the Quaternary glaciation, and using the names of local Alpine valleys in which evidences were found, they named the four periods of maximum ice extension, which had occurred in the Alps, the Gunz, Mindel, Reiss, and Wurm periods. Each of these is separated by a relatively mild interglacial period during which the ice retreated. They also established minor glacial phases at the fluctuating close of the Ice Age. This theory has been substantiated, in its entirety, or in part, wherever past glacial conditions have been examined; in particular the corroborative evidence of De Geer from his study of lake deposits should be noted.

With this problem of glaciation is associated that of climatic changes in the geological and historical past. The study of this problem is outside the sphere of geography. Its investigation must be left to the labours of the geologist and anthropologist; the geographer accepts their conclusions in so far as they are essential to the study of man's evolution in relation to the changing physical environment.

III

The comparison of features of the same character in different parts of the earth was the first obvious stage in establishing order out of the confusion of surface features. It was essential to the method of both Humboldt and Ritter, and had also been used by their predecessors. It was Ritter who first used the term Comparative Geography, and claimed it to be fundamental in geographical method. The concept, however, received only superficial application in his hands. He attempted to classify the major features of relief (see Chapter XV) and assumed the continents to be built on a uniform plan. He also compared and contrasted the configuration of the continents, to explain primarily the high stage of civilization in some parts of the world and its retardation in others. In fact, the comparative study of distributions, physical, climatic, biological and human, is the essence of modern geographic method. But it only becomes fruitful in physical geography when combined with causal interpretation—otherwise comparisons of form may be fortuitous and lead to no profitable conclusions.

The comparative method, with attempted explanations, is the basis of the general geography of Varenius. It is also adopted with excellent results by Humboldt. In physical geography he sought and found what he called 'analogies of form'. He applied the comparative method in his treatment of climate, in which he examines the distribution of temperature over the earth by using isotherms, and compares climatic conditions in the same latitudes, and accounts successfully for the differences. The same approach is the basis of his essays on the distribution of steppes and deserts, and distribution of plants. In fact, comparison and causation are the fundamentals of Humboldt's method. The comparative method requires maps for the accurate representation of distributions. Maps of Europe were prepared by Ritter early in his career, and Humboldt was responsible for the preparation of Berghaus's *Physical Atlas* (1838–42), the first atlas of its kind.

But the comparative method in physical geography did not begin with the two masters. Francis Bacon (1684) first drew attention to the tapering of the southern continents, and the same feature was remarked on by J. R. Forster in his *Bemerkungen auf einer Reise um die Welt* (1783). It was again expressed by Bergmann and elaborated by Kant, who showed that all the peninsulas with but a few exceptions point to the south. Kant also noticed that the trend of the east and west coasts of the Atlantic Ocean give it an outline comparable to that of a river, whence Humboldt produced his concept of the Atlantic Valley. Both writers also noted the parallelism of the Atlantic coasts between 10° N. and S., and the sympathy of curve of the projecting and reentrant angles of its opposite shores.

The Jesuit Kircher in his *Mundus Subterraneus* (1665) had visualized a world skeleton of mountain-ranges in both land and sea crossing each other at right angles, parallel to the lines of latitude and longitude. This idea was developed by the French geographer Buache in the middle of the eighteenth century, and was adopted by Bergmann and Kant, and even by Ritter in his early days. According to Buache, a mountain-system running from east to west, including the Pyrenees and the mountains of central Asia, completely engirdled the earth.

Such a suggestion could only be made before the existence of accurate maps, for as Reclus writes: 'We have only to glance at the maps at the present day ('sixties) to see how completely primitive this idea was as regards the harmony of the terrestrial configuration.' But the same idea, with a more scientific, though equally erroneous basis, lay behind De Beaumont's classification of mountains (1852).

The philosopher Krause in *Die Erde als Wohnort der Menschen* (1811) compares the configuration of the west coast of America and the east coast of Asia, the former concave and the latter convex, and he shows how this arrangement is repeated even in the small island fringes of eastern Asia. He also regarded the Atlantic as a midland ocean between the two great land masses, and Europe as a peninsula of Asia.

To Oscar Peschel (1826–75), professor of geography at Leipzig, is due the credit of laying the foundations of modern physical geography, through an elaboration of the comparative method as expounded by Ritter, with an attempt, however, to explain as well as classify surface features. His work in this connexion is published in his *Neue Probleme der vergleichenden Erdkunde als Versuch einer Morphologie der Erdoberfläche* (1870), and in two volumes of papers, edited after his death by Gustav Leipoldt, entitled *Physische Erdkunde* (1879).

In the former work Peschel criticizes Ritter for advocating a method in physical geography which he failed to apply. He argued that comparative geography should have a definite method and aim, like comparative morphology. The geographer should seek, with the aid of large-scale maps, similar physical features in different parts of the earth, compare their characteristics and origin, seek for transitional forms, and endeavour to relate them all genetically as in comparative anatomy. He also deprecates Ritter's teleological philosophy. 'An anatomist does not stop at the demonstration of homologies (of form); he may also attempt teleological interpretation when, for example, he demonstrates the functions of the skeleton and its component parts. But, in fact, he has then passed beyond the sphere of comparison.' That is, teleological treatment is beyond the sphere of comparative geography. For the 'geographical teleology' of Ritter, and his over-emphasis of the historical element, Peschel substitutes the comparative study of land-forms, and ignores their influence on human progress, which he considers beyond the scope of the subject.

Peschel's comparative method produced valuable results, and laid a basis for future study. His chief difficulty was the lack of data and knowledge, in his day, concerning the work of the agents of erosion—in particular running water—and of the details of regional geology and structure. By studying many topographic maps he sought 'homologies', or as Humboldt called them, 'analogies' of form, and then tried to trace their origins. While sometimes the method was successful, often fortuitous comparisons are made, and therefore reasons cannot be given for them. 'Seduced by the method of Cuvier, and forgetting that the evolution of terrestrial forms is governed by a principle different from that of organic forms, Peschel swamped comparative geography with innumerable problems' (Mehedenti).

An excellent illustration of Peschel's method is his essay on fiords in the *Neue Probleme*. From a study of topographic maps he determines the essential characteristics of fiords, 'deep and steep gorges in the coasts of continents and islands', which 'frequently extend inland perpendicularly or at a very high angle'. He notes the distribution of such coastal features, and concludes that the feature which 'strongly separates fiords from all similar coastal divisions is their local aggregation and gregarious occurrence'. They are found on mountainous west or north coasts in high latitudes. He concludes that they are 'empty dwellings of former ice streams', carved out of fissures due to earth movement. The same method Peschel successfully applied to other physical features, in particular lakes and islands (continental and oceanic). He also disproves the idea of the rectangular arrangement of mountain ranges. In discussing the origin of mountains he points out that all the young folded ranges are bounded on one side by land, usually high land, and on the other by an ocean deep, which has sometimes been filled by deposition—an idea taken from Dana's researches.

In an essay on 'Geographical Homologies' there are examples of the failure of his method. He notes homologies of form in the mountain framework of Borneo and the shape of Celebes and Halmahera, but for their origin he gives no satisfactory reason. Celebes, he says, is probably the skeleton of an old land-mass, but he cannot answer the question whether all three are three different forms, or the same one at different stages of development. Again, he notices that the islands of the Pacific are arranged in series like threaded pearls; and he notes the similar right-angle bends at the Gulf of Aden and Oman. The peninsulas and islands to the north of the continent, he notes, are oriented to the north, while on the east and west coasts they trend from north to south, and on the south they taper in the same direction. He discusses the similar shape of the three southern continents noted by previous writers and finds that though similar in shape, their forms—relief and drainage—differ. On this basis he concludes 'that the continents are older than their mountains', though this in the light of later knowledge is now completely disproved, for a continent is in fact a grouping of structural fragments.

Peschel established physical geography as a science. It had been neglected by Ritter, and Humboldt made no attempt to classify land-forms. Both merely dealt with the lands as a whole, and Ritter's classification, of a superficial kind, was merely based on relief. Peschel, however, attempted to classify, note the distribution, and explain the origin of specific land-forms, such as fiords, lakes, islands, valleys, &c.

His method of homologies failed largely because of the lack of knowledge of the agents of erosion. Equipped with the researches of the American geologists at the end of the century, and of the *Challenger* and other oceanic expeditions, and with subsequent advances in meteorology, Peschel could have treated his forms by dealing first with causes, and thus avoided his fortuitous comparisons. This development, however, was to wait yet another two decades.

Meanwhile, Elise Reclus had published *La Terre*, of which there is an English translation by A. H. Keane. The two main features of its physical geography are first the general treatment of the lands and oceans, their shape and major relief features; and second, the detailed descriptive study of land forms, e.g. lakes, deltas, glaciers, &c., illustrated with appropriate illustrations and maps. In his general treatment, Reclus reflects the comparative method of Ritter and Peschel. He writes as follows:

'The globe of our earth is in evident conformity to all the laws of harmony, both in the spherical uniformity of its shape and also in its constant and regular course through space. It would there-fore be incomprehensible if, on a planet so rhythmical in all its methods, the distribution of continents and seas had been accomplished, as it were, at random. It is true enough that the outlines of coasts and mountain ridges do not constitute a system of geometrical regularity; but this very variety is proof of a higher vitality and bears witness to the multiplicity of motives which have co-operated in the adornment of its surface.'

The contrast in land distribution between northern and southern hemispheres is noted. The three double continents form three parallel north-south series, each pyramidal in shape, and tapering to the south. This 'striking unity of plan, where at first sight all seems disorder and chaos', was a few years later interpreted in Green's Tetrahedral Theory (1875), to which previous reference has been made. Each pair of land masses is joined in low latitudes by an isthmus, and the

southern continents have smooth outlines as compared with
the northern. Two laws apparently determine the distribution of land and water—the old idea of geometrical arrangement still persisting—'one, according to which they are
arranged in circles obliquely to the equator, the other which
distributes them in three lines parallel to the meridian'. In
the Old World these two axes cross and produce the area of
greatest relief in the world. The southern tapering of the
continents, with the appearance of 'immense ruins', each with
a group of islands to the east of its extremity, is ascribed to
a deluge from the south-west, which dismembered the
southern continents, and carried the debris to the northern
hemisphere. Hence the reason for the great expanse and long
gradual slopes of the lands towards the Arctic.

The lands are divided into plains, plateaux, and mountains.
Plains are divided into two groups, mainly according to their
surface covering and their occurrence in similar latitudes.
(a) The zone of Landes, steppes, and tundra (Landes, polders,
N. Germany, pusztas of Hungary, the black earth belt, Caspian depression and Siberian steppes). (b) The zone of desert
plains, parallel to the first, including the Sahara, Arabia,
Iran, north-western India, and the Gobi. These two belts,
arranged parallel to the axes of the continents, are repeated in
the New World; here it is noted that the deserts occur on the
western side of both North and South America.

It is remarked of the plateaux that 'their height increases
in proportion to proximity to the equator as if the rotation of
the globe had caused not only the equatorial enlargement of
the planetary mass, but also the elevation of the continents
themselves'. The plateaux of central Asia are compared with
those of Europe in that both are bounded to the south by
mountain chains. In the chapters on mountains Reclus
accepts Elie de Beaumont's theory, and then proceeds to a
description of the principal systems of the earth.

Several works on physical geography appeared in England

between 1850 and 1880, which largely reflected Humboldt's teaching. The best were prepared by Mrs. Mary Somerville (1852) and D. T. Ansted (1871). The former is descriptive almost in its entirety, with no attempt to account for phenomena. Ansted, however, benefited from the researches of twenty years and produced a much better volume in which he deals with physical geography proper and with climatology. Chapters are devoted to valleys, plains, plateaux (plains over 600 ft. above sea-level), mountains, lakes, springs, glaciers, &c. There is, as in Reclus, no system of treatment of major earth features. Thus, plains, as designated, are of greatly varying morphological character, and include the llanos, pampas, selvas, the Sahara, &c., while in North America there are three types, distinguished by their vegetation, heathy or bushy plains, dry or rolling prairies, and wet or alluvial plains. An interesting feature is a chapter on the results of human agency, which is taken to include deforestation (with a discussion of its effects on climate), cultivation, drainage (polders), irrigation dyking, planting of sand-dunes, and the cutting of canals through isthmuses.

All these works dated before 1880, on the side of physical geography, are essentially empirical. The distinction between mountains, plateaux, and plains is purely arbitrary, and their grouping merely a matter of convenience. The vegetation of plains is as important a criterion as surface, in their subdivision. The obvious surface features—lakes, glaciers, springs, &c.—are described at length, but the underlying principles which determine their development and the use of these as a basis for classification are not yet appreciated. Finally the cycle of erosion has not yet been worked out, and genetic classification is therefore still impossible.

The development of land-forms through the work of running water on lands with different kinds of structure was worked out after 1870, and particularly in the 'eighties. While a number of reports and papers were published on the

subject during the period, Emmanuel de Margerie first published a book which dealt generally with river development
(*Formes des Terrains*, 1886). He definitely stated that land-
forms are the product of erosion, by atmospheric agents on
different 'structural surfaces'. The surface resulting from
erosion of the latter he calls the 'topographic surface'. Of
the agents of erosion, by far the most important is running
water, proof of which is given by (1) 'the complete drainage
of the lands, the proportion of volume of water to the size of
their channels, added to their ramifications, and the constant
correspondence of rivers with tributaries at their confluences,
and rivers with the sea at their mouths', (2) the lack of
harmony between the topographic and structural surfaces,
(3) the actual rate of erosion as determined by measurement.
The laws of river erosion are stated, and the development of
river systems is studied in regions of horizontal or slightly
folded, intensely folded, and faulted strata. The whole work
is illustrated by pictures, diagrams, and contoured maps,
many taken from the French 1/80,000 and 1/200,000.

De Margerie's work was followed by a number of others
during the last decade, each summing up the results of the
recent research, and offering various types of classification
to embrace all kinds of land-forms. De Lapparent's *Leçons
de géographie physique* (1886), James Geikie's *Earth Sculp-
ture* (1894), W. M. Davis's *Physiography* (1899) may be
specially noted—although the last published his researches in
article form in the 'eighties and after, some of which are contained in his volume of *Geographical Essays*.

The first comprehensive classifications were put forward by
Ferdinand von Richthofen (1886) and Albrecht Penck (1894).
Von Richthofen was the next great figure after Peschel in the
development of physical geography, though he was only one
of a growing number of professional geographers. He was
early attracted to geology, and after doing much work in the
Alps, he went out with a Prussian expedition to Eastern Asia

in 1860. He afterwards stayed in California for six years where he studied the relations between volcanicity and the occurrence and distribution of gold. In 1868, under the auspices of the Chamber of Commerce of Shanghai, he travelled China for four years. During this time, in addition to sending detailed reports to Shanghai, he carried on personal investigations into the coal deposits, geology, and structure of the parts of China he visited. On his return to Germany in 1872 he prepared his work for publication (*China: Ergebnisse eigener Reisen und darauf gegründeter Studien*, 1877; Atlas, 1885). In the first volume he discusses the structure of central Asia, and shows the influence of relief on the movement of peoplès, then puts forward his Aeolian theory of the origin of loess. In the second volume on North China he deals with geology, morphology, and the inhabitants and their activities. The third volume on South China and the atlas of China were published by a friend from his materials after his death.

As a physical geographer Richthofen was primarily interested in land-forms, and it is characteristic of the man as an explorer and an observer that his, the first, classification of land-forms should be contained in a book entitled *Führer für Forschungsreisende* (1886). The scheme is too elaborate to be given in its entirety and, moreover, it has since been considerably modified. Mountains, to take an example, he divides into six groups.

I. Tectonic mountains.
   (*a*) Block mountains (tilted block, flexure, and symmetrical block mountains).
   (*b*) Fold mountains.
II. Trunk or abraded mountains.
III. Eruptive mountains.
IV. Mountains of accumulation.
V. Plateaux (abraded plateaux, plains of marine erosion,

horizontally stratified table-lands, lava plains, river plains, plains of Aeolian formation).

VI. Mountains of erosion.

Coast lines he divided, according to the height and slope of the land, into cliff coasts, narrow beach coasts with cliffs, wide beach coasts with cliffs, low coasts, with sub-divisions of each according to whether the strike of the mountains is parallel, crosses, or is unrelated to the trend of the coast line.

In his *Morphologie der Erdoberfläche* (2 vols., 1894) Penck distinguishes six *topographic* forms, or *form elements*:

1. The plain or gently inclined uniform surface.

2. The scarp, or steeply inclined slope.

3. The valley, composed of two lateral slopes inclined to a narrow strip of plain which itself slopes down in the direction of its length.

4. Mountain, a surface falling away in every direction— which may be a point or a line (ridge).

5. Hollow, the converse of 4.

6. Cavern or space entirely surrounded by a land surface.

These forms do not occur alone, but are grouped together into *Landschaften* of different orders—districts, regions, and lands. The character of the topographic surface depends largely on the structural surface. The six chief *structural forms* are plains, with horizontal strata; slightly folded strata (*Verbiegungsland*); faulted blocks (*Schollenland*); intensely folded areas (*Faltungsland*); lava overflows (*Ergussland*); and intrusive volcanic masses (*Intrusivland*). With the action of the agents of erosion, and particularly running water, on the structural forms are produced the topographic or land-forms. On this basis Penck works out his classification.

Many works on various aspects of geomorphology have appeared since the publication of Penck's work. A revised edition of it was published in 1928, and it remains the standard work in German. The best single work covering the whole field of physical geography is De Martonne's *Traité de géographie physique* (abridged edition, English translation in 1926). Lake's *Physical Geography* is the best general English text-book.

IV

Oceanography is entirely a science of the nineteenth century. Some investigations were undertaken at the end of the eighteenth century mainly by Arctic explorers—Cook, Phipps, Scoresby, and especially John and James Clark Ross. But the foundations of the science were laid by Matthew F. Maury (1806–73), an American naval officer, who collected log records over a period of fifty years, and then published his results in *The Physical Geography of the Sea* (1855). This book dealt with the extent of the oceans, the forms of coast tides, ocean tides, and currents, the physical and chemical conditions of the sea and the organisms which inhabit them. With the help of three lines measured for the laying of transatlantic cables, he was able to sketch for the first time a section and map of the floor of the North Atlantic.

Oceanic records, however, are mainly due to the three simultaneous expeditions (1872–7) of the English *Challenger*, the German *Gazelle*, and the American *Tuscarora*. The *Gazelle* carried out investigations in the South Atlantic, Indian, and South Pacific oceans, and the *Tuscarora* in the North Pacific, to find a suitable course for a Pacific cable. The *Challenger* expedition, commissioned at the instigation of the Royal Society, crossed the Atlantic several times and the southern seas; it crossed the Antarctic circle and then proceeded to the North Pacific to Japanese waters, thence by Yokohama, Honolulu, and Tahiti to Valparaiso, and home by Cape Horn. The reports of the expedition were published in fifty volumes and form the basis of modern oceanography. Plumb-line soundings and deep-sea thermometer readings and samples of ocean sediments were collected from all latitudes. Thus was obtained our first accurate knowledge of the form of the ocean floors and the character of the deposits lying on them, of ocean depths, and of the nature of ocean life. Other similar expeditions were dispatched later to gather scientific data, and in 1902 there was founded an International Council for the study of the sea with its head-quarters at Copenhagen.

Following on these investigations, Sir John Murray (a member of the *Challenger* expedition) calculated the mean height of the land and the mean depth of the sea (2,250 ft. and 12,480 ft. respectively)—an elaboration of earlier esti-mates made by Humboldt, De Lapparent, and Wagner. Murray divided the earth into three zones, the continental area (dry land), the transitional area (submarine slopes down to 1,000 fathoms) and the abysmal area. Mill suggested the mean sphere level, which he calculated at 10,000 ft. below sea-level, as a more suitable boundary than the 1,000 fathom line. Hermann Wagner in 1894 amassed and criticized all the existing data, and recalculated volumes and mean heights. He reckoned the lands to cover 28·3 per cent. of the surface of the globe and the oceans 71·7 per cent., 2,300 ft. mean height of the land, 11,500 ft. the mean depth of the sea, and mean sphere level at 7,500 ft. below sea level.[1]

Wagner divided the earth into the following regions:

	Per cent. of Surface.	From (Feet).	To (Feet).
Depressed area	3	Deepest	−16,400
Oceanic plateau	54	−16,400	−7,400
Continental slope	9	−7,400	−660
Continental plateau	28	−660	3,000
Culminating area	6	3,300	Highest.

[1] All these calculations are approximate in view of unexplored lands in the polar regions and lack of accurate data for all the oceans.

In 1905 the Prince of Monaco used all the available data for the production of a bathymetrical map of the oceans, which is still the standard work on ocean depths.

V

The third branch of the physical basis of geography is climatology. Climate is the 'average condition of the atmosphere'; weather is 'a single occurrence, or event, in the series of conditions which make up the climate'. Meteorology is the physics of the atmosphere and is largely theoretical. Climatology or the science of climates is largely descriptive in that it 'aims to give as clear a picture as possible of the interaction of the various phenomena at any place on the earth's surface. It rests upon physics and geography, the latter being a very prominent factor' (R. de Courcy Ward, *Climate and Man*).

The raw material of climatology is therefore meteorology, and the development of the former is dependent on progress in meteorology, the collection of records, and methods of mapping such data. The collection of records and the production of weather charts began with Chevalier de Lamarck (1774–1829) who, with the co-operation of Laplace and Lavoisier, established observing stations and published a series of *Annuaires météorologiques* (1800–15). H. W. Brandes of Leipzig about 1820 compiled a series of daily weather maps based upon records for 1783, and later he published maps of the European storms of 1820, 1821, and 1823, and explained them as due to depressions moving from west to east. In 1825 J. P. Espy (1785–1860) pursued similar researches on thunderstorms and tornadoes, which were published in his *Philosophy of Storms* (1841) which established the 'thermal convection theory' of the origin of cyclones. In the 'forties, as government meteorologist, Espy prepared daily weather maps.

The development of the daily weather map was obviously dependent upon the daily collection of records from a wide area, and in the 'fifties, with the advent of the telegraph, the regular production of weather charts began. The Meteorological Office in London was established in 1854, and in 1860 Fitzroy, its head, began to collect daily reports and produced charts and daily forecasts from 1861. The Smithsonian Institution in the States began to publish daily reports in 1851. In 1858 the Paris Observatory began to issue an international daily bulletin, to which a daily map of isobars was added in 1863. All the civilized nations now produce daily weather charts and forecasts.

The first to offer an interpretation of the planetary wind system was Edmund Halley (1656–1742) who, in a paper entitled *An historical account of trade winds and monsoons observable in the seas between and near the tropics with an attempt to assign the physical cause of the said winds*, related the trade winds to the belt of rising air along the equator, and the alternation of monsoons in the Indian Ocean with the alternation of relative temperatures of land and sea. George Hadley (1735) accounted for the deflexion of the Trades, not by the movement of the sun and hence the point of maximum heat around the earth as suggested by Halley (and also Varenius), but by the earth's rotation.[1] These researches only applied to the trade wind belt. Their regular winds and the regular rhythm of the rainy seasons were far simpler to explain than the climatic conditions of higher latitudes. Here the variable strong winds, irregular storms and rainfall, distribution, in space and time, showed little relation to conditions in the tropics. Consequently down to the nineteenth century the climatic conditions of the westerly wind belts are often completely omitted in geographical treatises, e.g. Varenius and Pinkerton.

[1] This theory was re-invented independently and elaborated by John Dalton in 1834.

Advance began with the investigation of storms, tropical and extra-tropical, the collection of records from log-books, and the oceanic expeditions of the late nineteenth century. Maury, whose work on physical geography has already been mentioned, collected a vast number (over a million) of wind directions from log records. From these he was able to draw the following conclusions in his *Physical Geography of the Sea* (1855): 'In all latitudes between the parallel of 30° or 35° N., and the parallel of 30° or 35° S., the prevailing direction of the wind is from the eastward; in all other parts of the world, as far as observation has gone, it is from the westward.' The arrangement of the wind system he describes as follows: 'First, a low barometer near the Equator and a belt of calms; a high barometer with a belt of calms near each tropic; and again a low barometer in circumpolar latitudes, both north and south. And, secondly, from these two places of high barometer there is a general tendency of the air to flow, both north and south, towards the places of low barometer, and the direction of the currents in their tendency so to flow, is controlled by the influence of diurnal rotation.' In the 'fifties, Ferrel (1859) and Buys Ballot (1857) enunciated their laws relating to wind deflexion.

The investigations of Brandes and Espy into the characteristics and causes of storms in Europe and the West Indies, and Reid and Piddington's research on cyclones of the Indian Ocean, were followed by those of H. W. Dove, who in his *Law of Storms* (1852) stated that the general wind circulation consisted of an equatorial and polar current, and the changeable weather of the temperate zone was due to the conflict of these two currents. Espy in 1851 published his theory of the nature of cyclones from a series of synoptic charts, and from a study of West Indian tornadoes, and this theory was generally held till the close of the nineteenth century. His views may be summarized as follows (from Napier Shaw, *Manual of Meteorology*, vol. i):

1. The movement of the air is towards the centre.
2. A barometric depression in the centre.
3. A central ascensional air current.
4. The formation of cloud at a certain height and its radial dispersal accompanied by rain and hail after the cloud has reached some prodigious height.
5. The travel of the whole with the upper currents of the atmosphere.

In 1863 Galton, supporting Espy's theory, dealt similarly with areas of high pressure and called them anticyclones.

During the last twenty-five years great strides have been made in the study of cyclonic phenomena and a new theory has been established by the Norwegian, J. Bjerknes. This is usually known as the Polar Front theory, according to which, when two masses of air with widely differing properties of temperature and humidity come into juxtaposition, the energy set free on the front of contact generates disturbances which develop to form cyclones.

The condition of climatology in the middle of the nineteenth century is indicated by Reclus's treatment in *The Earth*. The idea of the planetary wind system is based on Dove's work. There are two great wind currents—a polar and equatorial, 'composed of masses of air flowing in opposite ways'. The poleward upper air current reaches the surface in middle latitudes and as a south-west wind increases in intensity, whereas the trade winds decrease in intensity as they approach the equator. The storms of the temperate belt are explained after Dove.

The description of rainfall distribution by Reclus is typical of the method adopted previously to recent research. The régime in the Trade Wind Belt is generally correct. Beyond, three rainfall regions are noted:

1. A belt with winter rains to the north of the limit of the trades where 'the aerial counter current moves south in

winter' (i.e. the Mediterranean winter rainfall regions, between lats. 30° and 40°).

2. Regions with spring and autumn rains which 'ought to comprise the countries over which the returning trade winds blow at the epoch when the sun is at the zenith of the Equator'; but the real causes, he says, are not yet proved.

3. In higher latitudes are summer rains (i.e. the continental areas in middle and high latitudes). 'This is because the sun, being then above the Tropic of Cancer, has brought back to the north the entire system of the trade winds and counter-trade winds; these latter therefore descend to the surface of the earth in high latitudes only, and there alone, in consequence of their conflict with the cold winds of the polar regions, is produced this notable increase of rain, owing to vapours brought from the tropics.'

Reclus attempts here to explain the seasonal and areal distribution of rainfall on general principles—though the latter are not yet fully developed, and the map of rainfall reproduced in *The Earth* is very crude. But it is more than is attempted in most books on physiography at the time, in which the rains are described in various parts of the earth, with no attempt to view the world as a whole.[1]

Fundamental to the development of climatology was the mapping of meteorological data. Humboldt first drew annual isotherms for the world in 1817, and these were later improved on by Kamtz (1832-6) and Mahlmann (1841). In 1852 Dove, in his *Die Verbreitung der Wärme auf der Oberfläche der Erde*, first drew maps to show mean monthly temperatures. Dove first used isanomalous temperature lines, and Krecke (1865) and Supan (1880) lines of equal tempera-

[1] A. Muhry (*Klimatographische Übersicht der Erde*, 1862) gave the first scheme of rainfall types. He took six belts in each hemisphere, bounded by latitude lines.

ture range. As long as it was supposed that the mean pressure at sea level was everywhere the same, no need was felt for isobaric maps, and this opinion was held till the middle of the century. In 1869 Buchan called attention to the need for taking account of differences in atmospheric pressure at sea-level in connexion with the barometric determination of heights. In the same year he published his treatise *The Mean Pressure of the Atmosphere and the Prevailing Winds over the Globe for the Month and for the Year*. As stated in the *Atlas of Meteorology*: 'This treatise did for pressure what Dove's great work did for temperature, but it was of far greater importance than Dove's famous work in stimulating research and in laying the foundations of scientific climatology.' Accurate and regular records of rainfall over the world— sufficient to produce a fairly accurate map—were not available till after the middle of the century. The first rainfall map appeared in the Berghaus Atlas (1845) on which isohyets, based on records, were shown only in Europe, the remainder of the world being shaded to show regions with heavy or little precipitation. The first isohyetal map of the world was prepared by Loomis of Yale University in 1882, and was revised in 1887 by Buchan. Later Supan and Herbertson prepared such maps based on more records, and Herbertson's is included in the *Atlas of Meteorology*.

Buchan in 1868 prepared charts of the world to show the distribution of temperature, winds, and pressure. These were greatly revised and improved in the volume of the *Challenger* expedition reports dealing with Meteorology. In 1887 Hann's *Atlas der Meteorologie* appeared as a section of the Berghaus Atlas. In 1899 was published Bartholomew's *Atlas of Meteorology* prepared by Buchan and Herbertson. This atlas gives, as stated in its introduction, 'in a clear and graphic form, the broad results of the science up to the present time, based on the patient and long-continued labours of myriads of observers in all parts of the world', and its outstanding feature 'is the comprehensive character of the subjects handled'. The maps are based on records from an aggregate of some 29,000 stations, and they 'summarize the observational data from the basis for the study of the climatologies of the globe'.

Climatology necessarily remained encyclopaedic in character until the general world principles of the planetary wind system and pressure and temperature distribution were known, and a division of the earth's surface, on the basis of definite climatic criteria, into regions with distinct types of climate, was devised. The classicists delimited the major zones torrid, temperate, and frigid, and this was the only division adopted until the end of the nineteenth century. These zones are based on sunshine; they are solar zones. A new delimitation, based on criteria of winds, rainfall, and temperature, was essential to geographers to afford a regional framework for the study of human activities.

The two first comprehensive classifications of climate, which form the basis of the majority of later ones, were prepared by W. Köppen (1884) and A. Supan (1903). Supan defined the world zones by isotherms—the hot belt by the mean annual isotherms of 68° F., the north and south cold caps bounded by the 50° F. isotherm for the warmest month, and between these the north and south temperate belts. Each of these belts is subdivided into climatic provinces, an aggregate of thirty-five. Köppen's classification is based on critical values of temperature and rainfall for the warmest or coldest month, or of the wettest or driest month for typical forms of plant life. He has five chief biological groups, controlled by temperature and rainfall.

A. Megatherms—plants requiring continuously high temperatures (over 18° C. all the year.)
B. Xerophytes—plants requiring dryness and high temperatures.

C. Mesotherms—plants needing moderate heat and moisture (some months over $18°$ C. but coolest month $>-3°$ C.).

D. Mikrotherms—plants needing less heat, cooler and shorter summers and colder winters (coolest month $<-3°$ C., warmest $>10°$ C.).

E. Hekistotherms—plants of the polar zones (all months $<10°$ C.).

Each of the zones is then again subdivided into regions, each named after a characteristic plant or animal.

Many other classifications have been prepared in recent years (see De Courcy Ward, *Climate and Man*), but the most important from the standpoint of the geographer is Herbertson's *Scheme of Major Natural Regions*. As this scheme was prepared by a geographer as a basis for geographical study, its consideration will be deferred (see Chapter XIX).

Climatology is thus a scientific product of the last forty years. The earliest work was naturally done in temperate latitudes in the United States and Europe. In tropical latitudes the monsoon climate of India was the subject of thorough investigation in the latter part of the nineteenth century, by H. F. Blanford, Sir Charles Eliot, and others. Blanford published his *Climate of India, Burma and Ceylon* in 1889, and Eliot edited the *Meteorological Atlas of India* at the beginning of this century (1906). Meteorological and climatological work throughout the world have been summed up in recent years in a number of books on climate. For several years Hann's *Handbuch der Klimatologie* was the only one of its kind in the field, but of works in English the standard compilation is Kendrew's *Climates of the Continents*, which is concerned exclusively with the classification, description, and causes of world climates.

Chapter XVII

THE DEVELOPMENT OF HUMAN GEOGRAPHY

THE influence of natural conditions on human activities and human mental and physical traits has been the subject of speculation throughout the ages, from the Greek philosophers onwards. Of the early moderns who were tempted by the problem mention may be made of the French Jean Bodin, who in the sixteenth century endeavoured 'to mark out on the surface of the earth the great forms in which human societies were inserted; frigid, temperate, and torrid zones with their subdivisions, eastern and western lands; plains, mountains, and valleys; barren lands or lands of promise; plains exposed to wind or protected from them. There was no rigidity, moreover, nothing tyrannical in the action of these fundamental geographical conditions on men. Bodin had a clear idea of the insufficiency and arbitrariness of a rigorous geographical determinism' (L. Febvre, *Geographical Introduction to History*, 1925), for he made allowance for the exercise of human and divine will. A century and a half later the Abbé Dubos wrote on the relation of the distribution of genius and ability in the arts and sciences, to the physiological effects of climatic conditions. Later, Montesquieu in his *Esprit des Lois* studies 'laws in general', and 'laws of civil slavery, menial service and the service of the state', and in their relations to climate he shows 'how the nature of the country influences those laws'. A century later (1861) Buffon's conception of the mutual relations of man and his environment shows a marked improvement, greater precision, and better understanding of the forces at work in those interrelations than his predecessors. 'For some thirty centuries the power of man has been joined to that of nature and has extended over the greater part of the earth. By his intelligence the animals have been tamed. . . . By his labours marshes have been drained, rivers embanked and provided with locks, forests cleared, moorlands cultivated. . . . The entire face of the earth bears to-day the imprint of man's power, which, although subordinate to that of nature, has often done more than she, or, at least, has so marvellously seconded her, that it is by our aid that she has developed to her full extent' (quoted by Febvre).

Henry Buckle in his *History of the Civilization of England* (1881), devoted over one hundred pages to the 'influence exercised by physical laws over the organization of society and over the characters of individuals'. He attributed individual and national character to the effects of physical conditions. Regions of great mountains or extensive plains (as in India) produce in man an overwrought imagination and gross superstition. When natural features are smaller and varied, as in Greece, reason develops in man at an early stage. Climate, he argues further, not only stimulates or debilitates, but also affects the constancy of man's work and his capacity for it. He considers 'that no people living in a very northern latitude have ever possessed that steady and unflinching industry for which the inhabitants of temperate regions are remarkable'. The reason, he asserts, is that the climate prohibits out-of-door employment for much of the year, resulting in desultory habits of work and fickleness of national character. This type of national character he notes in Norway, where the interruption in labour activities is due to the severe winter; and in Spain, where it is due to the heat and drought of summer.

Both Humboldt and Ritter appreciated the essential interdependence of man's activities and physical conditions, and both offered to their successors ideas which directed the lines of development of modern geography, either through their positive value, in which case they were accepted, or through their negative value, in which case they provoked further thought and gave birth to new and more precise

O

ideas. While Humboldt was definitely a physical geographer, Ritter set out to elaborate the human aspect of geography with special reference to the influence of physical conditions on the history of mankind. To posterity Ritter offered broad and fundamental concepts, but not a precise method of interpretation of the interrelations of man and nature. In his treatment geography was merely the handmaid of history.

Following upon the death of the two pioneers came reaction. Peschel, opposed to Ritter's views, and his neglect of the physical side, considered that geography consisted only of the study of the earth's surface features; man's activities lay beyond its scope. Thus was established that 'dualism' of geography, physical and human, which was characteristic of geography in Germany in the nineteenth century, and is still maintained by some. Peschel, however, though as a geographer confining himself to the earth's surface features, produced a work on ethnography—a description of the races and customs of mankind, though he made no attempt to relate human phenomena to environment. The geographer interested in these phenomena encroached upon all the kindred sciences, in his desire to weld together a systematic description of the races, languages, religions, social organization and cultures of mankind.

II

In the flood of thought following upon the theory of organic evolution, established and popularized by Darwin in his *Origin of Species*, in the second half of the nineteenth century, it was but natural that, with adaptation to environment through natural selection as the keystone to scientific thought, attempts should be made to evaluate and systematize the relations of man and nature, on the same principles as for other organic life. The bridging of the gap between the dual aspects of geography was effected in the latter half of the nineteenth century by two schools of thought, headed by Friedrich Ratzel in Germany and Frederic Leplay in France.

Friedrich Ratzel (1844–1904) began his researches in the natural sciences. He studied zoology and geology at the universities of Heidelberg, Jena, and Berlin, and in 1868 presented his doctorate thesis in zoology. He then worked at Montpellier for two years, investigating the zoology of the Mediterranean shores, and published his results in two volumes (1873–4). After serving in the Franco-German War, he became interested in journalism, and in the capacity of special correspondent for several papers he toured eastern Europe, Italy, and Sicily, and finally crossed the Atlantic to the United States, Mexico, and Cuba. It was his tour in the States which brought home to him the reality of geography. He subsequently published his researches and impressions of the States in two volumes, *Die Vereinigten Staaten von Nordamerika* (1878–80). On returning to Germany he became a *privat docent* in geography (1876), and in the following year was appointed geography master at the technical school at Munich. Here he stayed till 1883, when he succeeded Richthofen at Leipzig as professor of geography. Shortly before, in 1882, he had published the first volume of his *Anthropogeographie*. The second volume did not appear till ten years later (1891), though he produced the *Völkerkunde* (*History of Mankind*, translated into English) in the meanwhile.

During his eighteen years at Leipzig Ratzel exercised a great influence on the development of geography in Germany. He took a leading part, with M. Kirchoff of Halle, on the 'Central Committee for the Study of the Geography of Germany', and later he founded and edited the *Library of Geographical Manuals*, which includes works on climatology (by Hann), oceanography, glaciers, geodesy, mathematical geography, and botanical geography. His last work, published in 1897, was *Politische Geographie*.

Though there is no literal English translation of Ratzel's works on Anthropogeography and Political Geography, Ellen Churchill Semple has expounded his views in her two works on *Influences of Geographic Environment on the Basis of Ratzel's System of Anthropogeography* (1911) and *American History and its Geographic Conditions* (1913). There are several articles in the *Annales de Géographie* summarizing his views, Febvre subjects his whole system to searching criticism, and Brunhes outlines and discusses his contribution to Human Geography in a symposium on the *History and Prospects of the Social Sciences*, edited by H. E. Barnes (1925). It is from these sources that the following discussion is mainly drawn.

The influence of the new thought arising from the theory of organic evolution is evident throughout Ratzel's works. Man, like plants and animals, is a product of his environment, and his activities, development, and aspirations are ruthlessly conditioned by it. 'Man is a product of the earth's surface. This means not merely that he is a child of the earth, dust of her dust; but that the earth has mothered him, fed him, set him tasks, directed his thoughts, confronted him with difficulties that have strengthened his body and sharpened his wits, given him problems of navigation or irrigation, and at the same time whispered hints for their solution' (Semple). Regarding man as essentially passive, Ratzel sets out to establish the laws of the physical environment which determine human activities, distributions, and organization, in both space and time. With Euclidian precision, based upon cases which fit the argument, of which numerous examples are given to illustrate each 'axiom', he established a science, if such it may be called, which he named 'Anthropogeography', the science of the expansion and distribution of mankind over the earth, which has, as its fundamental concept, a rigid and fatal determination of man's existence, by the land on which he lives. But it must be admitted that in laying the foundations of a new science, the influence of one factor—the environment—will necessarily be apparently overestimated. For both writers are well aware of the complexities of their problem, of the rashness of broad generalizations, and the importance of psychological factors, 'easy to assert, but difficult to prove' (Semple). Unfortunately these considerations mentioned in the early pages nowhere mitigate their dogmatism.

Geographical influences on man are divided into four groups, direct physiological effects, psychological, the 'geographic conditions which influence the economic and social development of a people by the abundance, paucity, or general absence of the natural resources', and the influence of factors in directing the movements and ultimate distribution of mankind. Man and the State are entirely dependent upon them; no credit is given to human will and initiative; all is predetermined, and the soil is dominant, 'She has entered into his [man's] bone and tissue, into his mind and soul' (Semple); and 'always the same, and always situated at the same point in space, [the soil] serves as a fixed support to the changing aspirations of men'. It is this, says Ratzel, 'which governs the destinies of peoples with blind brutality' and 'a people should live on the land fate has given them; they should die there, submitting to the law!'

In his *Politische Geographie*, the geography of states, of commerce, and of war, Ratzel expounded the natural laws which govern the development and organization of states. He believed that 'society is the medium through which the State becomes attached to the soil' and hence the relations of society and the soil affect the State at every stage of development. We are then speedily led (Semple) to conclusions as follows—'The larger the amount of territory necessary for the support of a given number of people, . . . the lesser is the connexion between land and people, and the lower the type of social organization.' 'Every advance to a higher state of civilization has meant a progressive decrease in the amount of land necessary for the support of the individuals.' With the same stage of social development, 'the per capita amount of land decreases also from poorer to better endowed geographical districts, and with every invention which brings into use some natural resource'.

Ratzel regarded the State as an organism, in constant motion, expanding in area till reaching natural limits, and then eventually, and provided effective opposition is not offered by strong neighbours, overflowing those limits. 'Geographical, and still more political, expansion', according to Ratzel, 'have all the distinctive characteristics of a body in motion which expands and contracts alternately in regression and progression. The object of this movement is always the conquest of space with a view to the foundation of States, whether by nomad shepherds or by sedentary agriculturalists.'[1] Human groups and societies always develop within the limits of a natural framework (*Rahmen*), towards which from a small nucleus they expand and probably overreach; always occupy a definite location (*Stelle*) in the globe; and are always in need of sustenance. Hence their inevitable association with a definite area (*Raum*), which, with the increase of population, will inevitably expand, until met by natural or human obstacles. These are the three essential geographical facts which govern the character and progress of states.

As regards area Semple writes: 'A struggle for existence means a struggle for space', and hence the constant tendency for states to expand from a small to a large area. The Euclidean principle is here again plausibly set forth. Thus, 'we may lay down the *rule* that change in areal relations, both of the individual to his decreasing quota of land and of the State to its increasing quota of the earth's surface is an important

[1] Huckel, 'La Géographie de la circulation selon Ratzel', *Annales de Géog.* xv, 1906; xvi, 1907.

index of social and political evolution. Therefore, the rise
and decline of peoples and civilizations have depended on
their relation to area. Therefore, problems of area . . .
dominate all history.'

The wide area found by Darwin to be most favourable to
improved variation and rapid evolution in animals 'operates
in the same way in human development, and its influence
becomes a law of anthropogeography'. As in other organic
life, a large area guarantees racial and national permanence,
small area weakness and impermanence. In the process of
state expansion, 'gradation in area marks gradation in development'. Moreover, the higher the scale of civilization of
a community, the greater the density of its population, is an
axiom which is substantiated by a scheme of densities for
different modes of life, from industrial agglomerations to
hunters and pastoral nomads. Thus are expounded 'the laws
of the territorial groups of peoples and of states'.

Frontiers are essentially transitional zones. Political frontiers are subject to constant fluctuations due to the expansion
and contraction of the State. The breadth of such a frontier
zone is greatest when lying between a progressive and a backward underpeopled state, and narrowest between states of the
same order. Border communities, owing to their peripheral
location, are liable to unrest and a desire for political
autonomy.

In his *Anthropogeographie* Ratzel has three main objectives:

1. The mode of human distributions and groupings,
   ethnic, national, linguistic, religious, &c.
2. The interpretation of these distributions as determined
   by the physical environment.
3. The direct effects of environment on individuals and
   thus on society, e.g. the precise nature of the direct
   influence of climate on national character.

In the first volume (1882) Ratzel treats the second topic, i.e.

the causes of human distributions, i.e. the dynamic aspect of geography, and in the second, not published till ten years later, he deals with the facts of distribution, i.e. the static aspect of geography—a procedure which was criticized by reviewers, but justified in that the aim was to examine first the forces which govern human distribution; causes are sought for and the manner of the action explained. The first volume is an application of geography to history, and the second the geographical distribution of man.

Ratzel defines the limits of the *oekumene* or habitable land, and the uninhabitable lands within it, and he studies and attempts to account for the oscillations of their frontiers. On the edges of the habitable world are the border peoples, the outposts of civilization: Eskimos in the north, Hottentots, Bushmen, Australians, and Tasmanians in the south. He compares the respective locations of these peoples in the northern and southern hemispheres. The southern peoples lie between desert and habitable land and are gradually being pushed into the less favourable areas and exterminated, as contrasted to the former, where there is no competition with a more advanced civilization. The tapering of the southern continents results in greater ethnic variations than in the northern hemisphere.

Migrations within the *oekumene* are discussed in their relation to natural routes and barriers. The facts which govern man's distribution and development are treated. Climate determines the location of the chief centres of civilization in the temperate belt. Mountains function as frontiers and places of refuge, though rarely are they absolute barriers. Water bodies are one of the greatest obstacles to primitive man, and highways of intercourse when the art of navigation is mastered. Thus the Atlantic, long a complete barrier, is now to Europe and America what the Mediterranean was to Asia and Europe in antiquity. Here follows an analysis of the human geography of coastlines. Rivers and marshes prevent expansion, and the latter serve as places of refuge. Forests function similarly, and harbour backward people in their clearings.

By recasting Ratzel's work Semple deals with the human geography of the main types of environment, coasts, oceans and enclosed seas, waters, rivers, continents and their peninsulas, islands, plains (steppes and deserts), and mountains, with a final chapter, as in Ratzel, on the influence of climate on man. Throughout the book the same determinist attitude is adopted, with sweeping generalizations based on slender evidence. The general method of approach may be indicated by briefly summarizing the argument in the chapter on Island Peoples.

Owing to their small area and isolation the peculiarities of the flora and fauna of islands—poverty of species, endemic forms and survivals of primitive types—are reflected in man, though they are less marked owing to his greater mobility. Isolation results in homogeneity of race, language, and culture, a distinct national stamp, and conservatism. Freedom from disturbing contacts permits imported cultural strains to be brought to greater fruition (e.g. Japan) but 'rarely do they wholly originate the elements of civilization for their area is too small'. Owing to their isolation islands also serve as places of refuge and survival. They usually have a relatively dense population. This soon results in over-peopling owing to the small area. Hence, islands usually have a precocious form of scientific agriculture, or they are centres of emigration, or they resort to diverse methods (polyandry, infanticide) to keep down their numbers. This same condition also results in a low valuation of human life. Thus, cannibalism is asserted to be due to constant pressure on the limits of sustenance in the islands of Oceania. Elsewhere it is written that plains 'invite expansion, ethnic, commercial and political' and suffer from 'paucity of varied geographical conditions, and of resulting contrasts in their population'. These quotations, taken at random from Semple, serve to indicate the tenor of the deterministic concept.

In Ratzel's *Anthropogeographie* 'the whole life of men, all their multiple activities, human groups, and human societies are studied methodically, rationally, and collectively in relation to their geographical environment', with the object of restoring to geography 'the human element, the claims of which seem to have been forgotten and to reconstitute the unity of geographical science on a basis of nature and life. Such is a summary of the plan of Ratzel's work' (La Blache). He raised human geography to the dignity of a science—albeit his method was too rigidly scientific—which has since been critically examined in its scope, its method refined, and its outlook modified. The great contribution of the master to geography is summed up as follows by Raveneau:

'Between physical geography, sometimes predominant or exclusive, and the science of man, which neglects so easily the framework in which man moves and the space in which he lives, Ratzel has taken his stand. He has strongly insisted on the necessity for a broad view of general conditions, and grand laws on which depend the distribution of man over the earth. His principal merit is that he has reintegrated into geography the human element. By that he has given that science a new orientation and stimulus.'

III

Frederic Leplay (1806–82) made detailed observations of social and economic conditions throughout Europe, and established a new method in sociology. But his main conclusions were abused by his immediate successors, who, in their insistence on the rigid dependence of social organization on natural conditions, built up a doctrine as deterministic as that of the anthropogeographers.

Leplay started his career in the Ministry of Mines at Paris. He rapidly gained promotion, and in 1840 became Professor of

Metallurgy at the School of Mines, and later Chief Inspector of Mines. In this capacity he spent his long vacations touring Europe and Asia, studying social conditions of communities of all kinds. In 1870, after his retirement, he continued the same studies in France. As a result of these prolonged investigations he had accumulated details of the mode of life of three hundred families—based upon 'family budgets'. Thirty-six of these monographs were published in his *Ouvriers européens* (6 vols.) in 1855.

Leplay made two main contributions to sociology: first, his insistence on the treatment of the primary occupations, and secondly, the elaboration of his formula 'place, work, folk'. The primary occupations include such as hunting, herding, mining, fishing, &c. The type of environment in rural areas (place) must be related to the type of work, while upon the type of work largely depend the social organization and out-look of the folk. By means of his valley section, Leplay demonstrated the interdependence of place, work, and folk in the various natural environments (cf. Branford and Geddes, *The Coming Polity*). Rural surveys undertaken with this approach link geography, economics, and anthropology, the complex forming a comprehensive sociological survey. With specialization on environmental aspects, the study becomes social geography; if on aspects of work, it becomes economics; if on folk life and custom, anthropology. The sequence is revised in urban areas. Work is here a product of human contacts, and does not bear a direct relation to nature; cultural take the place of natural occupations. The immediate environment is a product of human endeavour —'art transforms place through purpose'. Hence the sequence in a civic survey becomes

$$\text{Polity} \longrightarrow \text{Culture} \longrightarrow \text{Art}$$
$$\text{(Folk)} \longleftarrow \text{(Work)} \longleftarrow \text{(Place).}$$

A combination of rural and civic survey gives a comprehensive regional survey. Such survey work is the prime aim of Leplay House in this country, and its method was elaborated by Sir Patrick Geddes and Victor Branford.

The geographer is peculiarly equipped for the execution

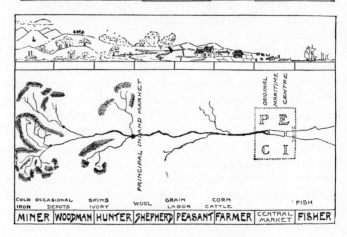

THE ASSOCIATION OF THE VALLEY PLAN WITH THE VALLEY SECTION

RURAL OCCUPATION & MARKET TOWN

FIG. 29. Leplay's Valley Plan.

of such comprehensive local surveys, and Leplay's scheme provides him with a method, and herein he may make valuable contributions to the common weal—through the study either of urban or of rural areas. The London School of Economics is at present undertaking a complete revision of Charles Booth's great work on *Life and Labour in London* inspired by the Leplay method. Many regional surveys, under the leadership of Prof. Patrick Abercrombie, have been undertaken for the guidance of future rural and urban development.

All these surveys are the work of specialists, but their results are of fundamental importance to the geographer, for they provide him with the raw material for his own type of synthetic investigation.

The principal disciples of Leplay in France were Henri de Tourville and Edmond Demoulins. These two writers dealt with the social life of typical communities, and claimed their organization and activities to be definitely determined by their natural environments. This method reaches its culmination in Demoulin's two-volume work on *Comment la route crée le type social*. He deals with the chief social groups throughout the world, and with Ratzelian dogmatism ascribes all their traits to the fatalistic dictates of the routes which they followed in their early migrations. The first volume, dealing with the routes of antiquity, gives chapters on the steppe route and the Tartar-Mongol type associated with it, and the invasions of nomadic pastoralists. With the tundra and savana are associated the Lapps and Eskimos and Red Indians respectively. The second volume deals with modern routes—the original fiord and later Saxon plain environments of the Franks and Saxons; the Celtic migration and their social organization; and the spread of the Slavs—these are a few random examples. In his Introduction Demoulins dogmatically declares his thesis. 'The primary and decisive cause of the diversity of peoples and races is the route which they have followed.' 'It is the route which creates both the race and the social type.' But he goes farther than this, for 'if the history of man began again, without any changes on the earth's surface, it would repeat itself in its main features'.

IV

Ratzel's human geography met with protest and criticism on account of its materialistic basis, and the wide scope, encroaching on the social sciences, which he claimed for it.

The new school of French sociologists were up in arms. Durkheim, in a review of *Anthropogeographie*, complains of its indeterminate object and method. 'It amounts, in short, to studying all the influences exerted by the environment on social life in general.' These facts, he argues, are too diverse to be included in one science; a single person cannot possibly master all the problems which they raise. Ratzel, continues Durkheim, overemphasizes environment at the expense of contributory social factors, which modify the human response.

The anthropologists also were in opposition. Ratzel glibly ascribes physical and mental traits to the influence of environment, and dogmatically discusses problems which are still unsolved by psychologists and anthropologists, notably the respective parts played by environment and race, in human mental and physical make-up. 'The full meaning of life', writes Marett (*Anthropology*), 'can never be expressed in terms of its material conditions'. The schools of Ratzel and Leplay, he continues, are fertile in generalizations that are 'far too pretty to be true'. Man cannot be regarded as putty in the hands of Nature. He is a rational being, with a social heritage, and his actions are not directly prompted by his surroundings. Otherwise, says Marett, 'Why do men herd cattle, instead of the cattle herding the men?'

The modern concept of human geography, which has been labelled Possibilism, as opposed to the Ratzelian Determinism, had its most brilliant exponents in Paul Vidal de la Blache, who died in 1918, and Jean Brunhes, who died in 1930. Vidal de la Blache first turned to geography in the early 'seventies, studied the works of Humboldt and Ritter, and travelled widely in Europe. He taught at the École Normale Supérieure for twenty years (1877–98), and then at the Sorbonne until his death, where he exercised a great influence over the thought and research of a following of enthusiastic pupils. He founded and published a number of brilliant articles in the *Annales de Géographie*, and an *Atlas of Historical Geography* in 1894; and some of his lectures on human geography were published shortly after his death under the title of *Principles of Human Geography*. La Blache realized the necessity for detailed synthetic studies in geography, and through his inspiration and direction a number of regional monographs were published,[1] while he wrote his *Tableau de la géographie de la France* in 1903, still the classic exposition of regional method.

His main aim in geography was to study cause and effect in related phenomena, and then to co-ordinate and establish general principles through the comparative study of different parts of the globe, on the lines suggested by Humboldt and Ritter.[2] 'The field of study *par excellence* of geography is the surface, that is, the ensemble of phenomena which lie in the zone of contact between the solid, liquid, and gaseous bodies which form the earth.' All these phenomena are to be interrelated on a regional basis, compared and correlated in their world distribution. On the side of the physical environment, the geographer must borrow freely from the geologist, for 'present forms are unintelligible unless their past evolution is known', while in seeking the relations of man with his environment, a number of factors, other than the direct effects of environment, take effect. Man is an active agent; all human facts cannot be explained through environmental control. Response is conditioned largely by aptitude, stage of development, and social heritage. Hence the necessity for a thorough grounding in geology and history. 'The bridge between geology and history is formed by geography. That which geography, in exchange for the help it receives from other sciences, can offer to the common fund, is the aptitude of not breaking up that which nature has brought together, of understanding the correspondence and correlation of facts,

---

[1] He also wrote a regional monograph himself, *La France de l'est*.
[2] See Introduction to *Atlas of Historical Geography* (1894).

either in the terrestrial environment which surrounds us, or in the regional environments where they are localized.'

The modern concept shifts the centre of gravity from nature to man, the active agent. His mode of life is the product, not of the dictates of environment, but of a complex of factors: social, historical, and psychological. As La Blache pointed out, 'force of habit plays a great part in the social nature of man', and according to the social complex, so, with similar environments, may be associated different modes of life. The environment contains a number of possibilities, and their utilization is dependent entirely on human selection. A social complex, engrained through habit, may therefore result in the neglect of certain possibilities. In short, 'the outstanding psychological fact, then, is the antithesis of a rigid fatalistic determination of human acts by climate and soil. It is this—that natural surroundings whether as a whole or in detail, react on us just so far, and just in such a way, as we adopt them; in other words, according to our interpretation of them' (Brunhes, *Human Geography*). 'The man is there, the flint is there, but it is the man who makes the spark fly,' for, 'all the essential facts begin and end in facts of psychology'.

Febvre, to whom the label 'Possibilism' is due, again clearly states the attitude. 'It is not true that four or five geographic influences weigh on historic bodies with a rigid and uniform influence; but at every instant and in all phases of their existence, through the exceedingly supple and persistent mediation of those living beings endorsed with initiative, called men, isolated or in groups, there are constant, durable, manifold, and at times contradictory influences exercised by all those forces of soil, climate, vegetation—and many other forms besides—which constitute a natural environment.' A final sentence again from Febvre may serve as a byword for modern geographic method. 'There are no necessities, but everywhere possibilities; and man, as master of the possibilities is the judge of their use. This, by the reversal which it involves, puts man in the first place—man, and no longer the earth, nor the influence of climate, nor the determinant condition of localities.'

Jean Brunhes, the most brilliant disciple of Vidal de la Blache in the field of human geography, was appointed to the first chair of human geography established in either Europe or the United States, at Lausanne in 1907. A similar post was created for him at the Collège de France at Paris in 1912, and he held this until his death in 1930. La Blache's philosophy of geography is the basis also of Brunhes's work, and its fundamentals are summarized by the latter as follows. 'In recalling the concrete idea of the physiognomy of the earth as modified by man, in perceiving not only an intervention of man in the equilibrium of inorganic nature but also that class of relations which places men at odds and in competition with the other living beings, in studying the human facts only in their relation with the surface from which there develop these multiple actions, incessantly repeated, in utilizing the method of the biological sciences, in using, likewise, perfected instruments for his work (exact maps, results of exploration, verification of data), in taking account, as a starting-point, of the general facts of distribution, and in arriving at a sort of "ecology" or conception of the complex bonds existing between a country and its inhabitants, without neglecting the influences from elsewhere—in all these points this doctrine is characterized.' (See article by Brunhes on 'Human Geography' in *History and Prospects of Social Sciences*, ed. by H. E. Barnes.)

Brunhes narrowed down the field of human geography as conceived by Ratzel, by taking as a basis the evidences of man's use and occupation of the environment. Ratzel had dealt in volume ii of his *Anthropogeographie* with the 'works of man on the Earth's surface', and it was this aspect which Brunhes elaborated. He introduced order and a principle of

classification and a method of treatment into the facts of
human occupation, ranging from the vital necessities of food,
clothing, and shelter, to the most complicated social, political,
and economic facts. The essentials of human geography are
the evidences of human action on the earth's surface. These
facts he divides into six types or three pairs:

(1) Facts of unproductive occupation of the soil—houses
and roads.
(2) Facts of vegetable and animal conquest.
(3) Facts of destructive economy in the vegetable, animal,
and mineral kingdoms.

This scheme is the framework of Brunhes's work on *Human
Geography*, which, based on the 'essential facts', covers the
whole field of social, economic, and political geography.

Professor P. M. Roxby summarized the modern concept of
human geography in his presidential address to the Geography Section of the British Association in 1930. He also
defined the scope and aim of human geography. In his view,
it consists of '(*a*) the adjustment of human groups to their
physical environment, including the analysis of their regional
experience, and of (*b*) inter-regional relations as conditioned
by the several adjustments and geographical orientation of
the groups living within the respective regions. The term
"adjustment" I take to cover not only the "control" which the
physical environment exerts on their activities, but the use
which they make and can make of it. Human geography is
the study of an interaction rather than of a control. The
adjustment has distinct but usually closely related aspects
which form the main branches of human geography'.

These four principal aspects of human geography are
racial, social, economic, and political. The first treats the
distribution of racial types, and their mental and physical
traits, relatively to the environments in which they are found,
and in so far as they affect the human response in different

environments. In the light of what he can glean from anthropology the geographer should study 'the relative aptitude and adaptability, climatic and otherwise, of various racial groups for developing them (regions) and the extent and manner in which co-operation between different groups may, in certain cases, be secured for this end'. Economic geography includes the geography of production and consumption, exchange and transport. Social geography is defined as 'the analysis of the regional distribution and interrelation of different forms of social organization arising out of particular modes of life, which themselves represent a direct response—although we may concede to M. Febvre not necessarily the only possible response—to distinctive types of physical environment'. The function of political geography is 'to study and appraise the significance of political and administrative units in relation to all the major geographical groupings, whether physical, ethnographic, social and economic, which affect mankind.' Finally 'historical geography is essentially human geography in its evolutionary aspects. It is concerned with the evolution of the relations of human groups to their physical environment and with the development of interregional relations as conditioned by geographical circumstances'.

Barrows, in an article in the *Annals of the Association of American Geographers* (1927), borrows a botanical term, and calls geography 'the science of human ecology', the aim of which is not to examine the character and occurrence of features of the environment, but to examine human responses to them. Man is the central theme, and all other phenomena are interpreted only in so far as they are related to man's adjustment to them. In relation to the social sciences, in which geography is included, sociology deals with types of social organization, and the study of social life, with a view to the guidance of social service; economics explain the relations between men in earning their livelihood; history deals with human time relations; geography with human place relations. The last two Barrows considers are to each other as chronology is to ecology.

This concept of human geography, however, has been subjected to criticism in recent years by a number of continental and American geographers. The most able exposition of the new concept and technique is contained in the works of Siegfried Passarge of the University of Hamburg, while an excellent summary of recent continental views, and a keen criticism on the basis of scientific method, has been published by Carl O. Sauer of the University of California.[1] According to this view geography is primarily concerned with the description and interpretation of the earth's features, physical and cultural. It aims to describe systematically the physical and cultural landscapes,[2] and to interpret these, in relation to all contributory factors, whether they be environmental or no. It is not, on this view, concerned with the investigation of human adjustment to the environment, but with the description and interpretation of areal occupance.[3]

Each of these branches of human geography is of recent growth, and its development is intimately associated with that of the cognate sciences concerned in each case. The scientific study of races began with J. F. Blumenbach in 1781, and was carried on by Prichard, Peschel, Ratzel, Huxley, Keane, Deniker, Quatrefages, Topinard, and others in the nineteenth century. But for long there was no definite conception of the meaning of race, and it was usually confused with culture and language. The progress of anthropology was entirely dependent on measurement of human physical traits, and on this basis, at the end of the century, there appeared authoritative works on racial types by Sergi (*The Mediterranean*

[1] Carl O. Sauer, 'Morphology of Landscape', *Univ. of California Publications in Geography*.

[2] See P. Bryan, 'Cultural Landscapes', in *Geography*, 1932.

[3] See articles by Whittlesey, Wellington Jones, and others in *Annals Assoc. American Geographers*.

Race), Beddoe (*The Races of Britain*), and Ripley (*The Races of Europe*). With the collection of records the classification of races of mankind has been attempted in recent years, on the basis of various criteria, by different authorities—Haddon, Fleure, Dixon, &c.

Geography, according to Roxby's definition, is concerned with the influence of environment on the evolution of races, a problem which at present is largely speculative, and the solution of which lies with the qualified archaeologist, anthropologist, and physiologist. With this aspect are associated the researches of Professor H. J. Fleure and H. J. E. Peake on racial and cultural geography in their evolutionary aspects, in relation to environment. A second problem of vital importance to the geographer is—how far do the mental traits of racial types influence their response to environment? How far are mental traits racial in character? How far are they hereditary or acquired through the social environment? Here again are unsolved problems, still speculative. Their solution depends on the experimental psychologist and, as opposed to recent schools of thought, should not be biased by political prejudice. In this connexion we may note in particular the work of Dr. William Macdougall, the psychologist, on *The Group Mind*, Hertz's recent work, an unbiased discussion of the whole problem, on *Race and Civilization*, and the quantitative researches of Ellsworth Huntington on the influences of weather variations on human efficiency, mental and physical (see, for example, his *Civilization and Climate*, 1915).

Through his concept of 'possibilism', or human adaptation to environment, Vidal de la Blache gave social geography a sound foundation. On the basis of the primary human needs he showed how the nature of the food supply, clothing, and shelter is directly due to adaptation to environmental conditions, except in the great urban agglomerations. He described the distribution and density of population of the lands,[1] and investigated the environmental conditions which favoured the growth of the great agglomerations, the limits of which he clearly defines. Mill and Dickson (*Scottish Geographical Magazine*), at the opening of the century, were attracted by the problem of the distribution and growth of world population, and Close and Fawcett have dealt with it quantitatively in recent years (see *Sociological Review*, 1925, and *Geography*, 1927). In two brilliant articles in the *Annales de Géographie*,[2] La Blache develops fully his concept of modes of life ('genres de vie') as adaptations to varying environments. In recent years the social geography of various communities has been investigated (e.g. the geography of pastoral life), but perhaps the best study of the kind, though now old, is De Preville's *Des Sociétés africaines*. R. U. Sayce's social and ethnographic studies of the Bantus (*Geographical Teacher*) should be particularly noted, while Ogilvie has suggested, in the same journal, methods of correlating the seasonal rhythm of activities in primitive communities with climatic and other conditions.

The distribution and form of settlements in their geographical relations is an important aspect of social geography. Meitzen, in his classic work *Siedelung und Agrarwesen* (1895) ascribes the two main forms of rural settlement in north-western Europe, nucleated and dispersed, to Germanic and Celtic origins respectively. It is now agreed that the form and distribution of rural settlements are due to a number of factors, racial, geographical, economic, &c. Aurousseau has submitted articles on types and classification of rural and urban settlements in the *American Geographical Review*, and the present condition of the geography of rural settlements is summed up in two excellent articles by Demangeon in the

[1] P. Vidal de la Blache, 'La Répartition des hommes sur le globe', *Ann. de Géog.* xxvi, 1917.

[2] P. Vidal de la Blache, 'Les Genres de vie dans la géographie humaine', *Ann. de Géog.* xx, 1911.

Annales de Géographie. The great importance of the problem is shown by the existence and publications of a Commission on Types of Rural Habitation established by the International Geographical Union.

Urban geography, in some respects an independent branch of human geography, deals with the site and location, plan, development, and functions of towns. Many works have been published on the geography of individual towns,[1] and problems investigated in recent years include the geographical interpretation of distinct regions within urban agglomerations (e.g. Roxby's 'Merseyside' in *Geography*; De Geer's 'Greater Stockholm' in the *American Geographical Review*, &c.), and the delimitation and characteristics of the zones of influence of large cities (Dickinson in *Geography*, 1930). But as yet there is no agreement as to what constitutes a town, or as to the general functional classification of towns (*id.*, 1932). If human geography is concerned with the description and interpretation of the cultural landscape, then 'urban morphology', the description and interpretation of urban plans, is an important branch of the subject. Yet it has been neglected. The few thorough contributions which have been made have come almost exclusively from German geographers. Here may be noted in particular W. Geisler's 'Die deutsche Stadt', Rudolf Martigny's 'Die Grundrissgestaltung der deutschen Siedlungen' (*Pettermanns Mitteilungen*, 1928), and a symposium, edited by Passarge, on *Die Stadtlandschaften* (1930). Economic geography, owing to its utilitarian character, has received more attention than any other branch of human geography. It is the only aspect of geography in which Britain can claim distinction, for George Chisholm's *Commercial Geography*, first published in 1889, is still the standard work in English. In the United States, recent de-

---

[1] e.g. R. Blanchard's *Grenoble* (1911) and J. Levainville's *Rouen* (1913); and Hans Robek's 'Innsbruck' (*Forschungen zur deutschen Landes- und Volkskunde*, 1925).

velopment, and the need for the careful exploitation and conservation of its natural resources, have been responsible for considerable attention to the geography of production. It owed its chief stimulus to the Wharton School of Commerce and Finance after 1900, under the leadership of J. Russell Smith. To-day there are many economic geography specialists in the States, and a special journal, *Economic Geography*, is devoted to the publication of research in this sphere. In Germany, till the end of the nineteenth century, attention was mainly devoted to physical geography and anthropogeography, but Andree edited an exhaustive *Geographie des Welthandels* in three volumes between 1862 and 1877, which is at present being re-edited with Dietrich, Hassert, Leiter, and Sieger as its chief contributors.

Political geography was first fully treated by Ratzel in his *Politische Geographie* (1879), a fourth edition of which was published in 1924, and later by A. Supan in *Leitlinien der allgemeinen politischen Geographie* (1922), and Rudolf Kjellen in *Der Staat als Lebensform* (4th ed., 1924). A German periodical, *Geopolitik*, edited by K. Haushofer, is devoted exclusively to political geography as established by the above authors. Political geography has two aspects, first, the relation of the State to all geographical distributions and factors, and second, frontiers. With regard to the first, its study was fostered by the revision of the political map of Europe by the peace treaties following the Great War. Many articles were published in various periodicals on the relations of the new and old political frontiers to linguistic, racial, and economic distributions. As an instance may be quoted Unstead's paper on the 'Belt of Political Change in Europe' (*Scottish Geog. Mag.*)—i.e. the belt of new states stretching from Finland southwards to the Balkans. Newbigin and Fleure have also published geographical studies of the political changes wrought by the Treaty of Versailles (*Aftermath*, and *The Treaty Settlement of Europe*); but the standard work of this kind is Isaiah Bowman's *New World*. Outside Europe Roxby has made two brilliant contributions to the political geography of the East, in his 'Far Eastern Question in its Geographical Setting' (*Geog. Teacher*). Many works have been published on the geography of frontiers, but probably L. Dominican's *Frontiers of Language and Nationality in Europe* (1926) is the best of its kind. Political geography is defined by Brunhes as the 'general and synthetic study of the geographical conditions of the development of political societies'—that is, it deals with the geographical factors and conditions of states, their territory, roads and frontiers, and capital cities. It is merely a part of the role played by geographical factors in history, and he therefore suggests that a more appropriate term would be the 'Geography of History'. The subject thus defined was treated by Camille Vallaux in two volumes, *La Mer* and *Le Sol et l'État*, in which he formulates the fundamental problems of the geography of· political societies. At a later date (1921), Brunhes and Vallaux produced in collaboration their *Géographie de l'histoire, Géographie de la paix et de la guerre sur terre et sur mer*, in which the whole field of political geography, based on the concept of possibilism, is expounded.

Historical geography was defined above (Roxby) as human geography in its evolutionary aspects. This definition obviously gives virtually no limit to its scope, for racial, social, economic, and political geography may all be approached historically. On the other hand, a second method of approach is through the study of the changing forms of the cultural landscape within a distinct regional unit. It is possibly owing to its enormous scope, that so little specific historical geography has been attempted of recent years. As a matter of fact, there is still considerable difference of opinion regarding the nature and scope of historical geography—though the difference is mainly one of terminology.

The history of geography, such as is being attempted in these pages, is the history of the development of geography as an organized body of knowledge—it is quite distinct from, though its progress is obviously dependent on, the geography of exploration and discovery. Historical geography is defined by Brunhes as 'the study of the regional development of the earth's surface either from the point of view of its physical conditions, or from the point of view of the transformation of the political or administrative organization'. Contributions in this field are A. Himly's *Histoire de la formation territoriale des états de l'Europe centrale* (1894), Bodo Knull's *Historische Geographie Deutschlands im Mittelalter* (1903), and K. Kretschmer's *Historische Geographie von Mitteleuropa* (1904), and Freeman's *Historical Geography of Europe*. The first of these works, by Himly, 'examines and analyses the transformation of the political organization of a part of Europe, exactly as one would examine and analyse the transformation of the landscape or population of a region' (Brunhes). (See above for his definition of the geography of history.)

Hereford B. George was one of the earliest exponents of historical geography in England. His *Relations of History and Geography* (1901) was the first book of its kind in English. Miss Semple's work on *American History* contains the ideal concept of historical geography, but is marred by its Ratzelian determinism. The best example in English of the modern method in historical geography is contained in Roxby's work noted in the previous paragraph. Vaughan Cornish in his *Great Capitals* deals with the location of capital cities throughout history in relation to the frontiers of their states and the accessibility of 'bases' with adequate food supplies.

Chapter XVIII

THE DEVELOPMENT OF BIOGEOGRAPHY

BIOGEOGRAPHY is the study of the relation of plant and animal life to the physical environment. Its two branches are zoogeography or the geography of animals, and phytogeography or the geography of plants. It is essential at the outset to fix clearly the exact relation of these two studies to human geography as defined in the last chapter. It has been repeatedly stated that human geography is concerned with the interrelations of man and nature; that it is not concerned with the distribution of all phenomena on the earth's surface, but only of those phenomena which bear some relation to man. All such features may be regarded as comprising the physical environment. The terrestrial covering of vegetation is thus a fundamental environmental factor, and the study of vegetation in its relations to man is a vital part of the physical basis of geography.

The geography of animals, however, is of little direct consequence to human geography. Zoogeography is geography in so far as it deals with distributions, and with the adaptation of animal life to environment. But of what consequence to a study of the relations of man to the earth is the distribution of wingless birds, crustacea, and mollusca, of the adaptation in colour and form of various forms of animal life? Moreover, given the distribution, its interpretation depends as much upon migrations in the geological past and on the investigation of fossilized fauna, as upon the effects of environment. And adequate treatment depends upon highly specialized knowledge. But facts of animal distribution which concern man should by all means be included by the geographer in his scheme. He may draw from the zoologist facts of zoological distributions to assist the complete interpretation of the physical environment in its relations to man;

but all other aspects of animal distribution, in relation to environment and migrations, lie beyond the sphere of human geography.

Thus, to illustrate the point, it is significant to the geographer that the principal seats of mammalian development are in the Old World of Europe and northern Asia. Here were most of the animals suitable for domestication, for example, the horse and sheep, whereas the southern continents are poor in mammals, which are smaller and weaker than those of the northern hemisphere, and have peculiar characteristics rendering them unsuitable for domestication— the marsupials of Australia, the antelopes of Africa, and the llama and guanaco of South America. Into each of these regions domesticated animals have been imported from the northern hemisphere. Two of the best examples of such introduction are the cases of the sheep and rabbit in Australia, the present stock of sheep having its beginnings in Capt. McArthur's little flock, while the rabbit has bred at such a prodigious rate that it is now a national pest, a source of public danger and considerable private profit as an export. These factors of mammalian distribution all have great human significance, but it must be sufficient for the geographer to accept the facts of distribution from the zoologist. It is not part of his task to investigate the details of faunal distributions and origins.

Zoogeography, or the distribution of animals, and plant geography, or plant ecology, as it has come to be called in recent years, are both products of the second half of the nineteenth century, as offsprings of zoology and botany, nurtured by the doctrine of organic evolution. Great advances in the systematic study of organic life were made in the second half of the nineteenth century. Buffon (1707–85), the French naturalist, published his monumental work on *Natural History* in forty-four volumes (1749–1804), and Linnæus (1707–78), a Swede, introduced system into the classification of plants through comparative study, and his labours were published in 1735 in his *Systema Naturae*. The former excelled in description; the latter in empirical classification, and both held to the idea of the fixity of species.

In the first half of the nineteenth century further progress was due partly to Bonnet (1720–93), who, in his *Contemplation de la Nature* (1764) revived the Greek idea of the gradual development of life, but mainly to three French scientists, Lamarck (1744–1829), who published his views on the inheritance of acquired traits as an explanation of progressive evolution, Cuvier (1769–1832) who, in his systematic zoology clung firmly to the theory of special creation, and St. Hilaire (1772–1844), also a comparative anatomist, who maintained the similarity of structure or 'homology' of all living beings, a view which has since been fully worked out and confirmed. In the botanical field a pupil of Linnæus, B. de Jussieu (1699–1767), turning from his master's system of empirical classification, prepared a system based on resemblances between plants, and the association of related forms. This work was carried on with brilliant results by A. de Candolle (1778–1841), and now forms the basis of the present system of botanical classification.

I

Considerable advances in the reasoned distribution of plants and animals were made in the early decades of last century. German scientists directed their attention to the climatic limits of trees in northern Europe and to the upper limit of tree growth in the Alps. At first correlations were sought with mean annual temperature by Leopold von Buch and later Georg Wahlenberg, as a result of his travels in Lappland (1800–10), Switzerland (1812), and the Carpathians (1813) (where he studied the vertical belting of vegetation), concluded that winter temperatures are critical for true growth. The duration of light was later considered by Boussingault (1844) by a comparison of plant growth in high tropical altitudes and in central Europe. De Candolle studied the influence of climatic conditions on plants in his *Essai élémentaire de géographie botanique* (1820), enlarged in *La Géographie botanique raisonnée* (1855). He showed how plants with resins and hard skins can tolerate cold winters; why Alpine plants, requiring much light and little warmth, are stunted in low hot plains; how the moisture requirements of a plant increase with the increase of its foliage, so that trees with needle-shaped leaves are able to withstand drought. The results of all these early works on plant distributions and forms were summed up in Schouw's *Grundzüge einer allgemeinen Pflanzengeographie* (1823), with an atlas showing the distribution of certain plants and their climatic limits. As early as 1806 Carl Ritter in a small atlas of Europe showed the distribution of forest and cultivated land in six belts, and the polar limit of evergreen trees and shrubs in latitude 47° N.

In 1805 Humboldt in collaboration with Bonpland published his *Essai sur la géographie des plantes*, and in 1808 *Ansichten der Natur* (Engl. trans. *Aspects of Nature*, by Sabine). The evolutionary aspect of plant development and the theory of plant dispersion had not yet been formulated, and these early efforts attempted to relate plant distributions, individuals, and associations, with climatic conditions. Humboldt, with true geographical perspective, regarded the vegetation covering as a whole, and he recognized large zones, each of which, with 'its own peculiar advantages, has its own distinctive character'. He recognized the regional character of the distribution of vegetation, for 'each region of the earth has a natural physiognomy peculiar to itself'. With the distribution of individual plants he was not concerned—'the botanical specialist divides many groups which the physiognomist is obliged to unite'.

From his detailed investigations in the New World between

latitude 60° N. and 12° S., in Europe and in central Asia, he recognized sixteen different forms which determine 'the aspect or physiognomy of nature'. Each of these is the dominant member of a particular association adapted to local conditions. Thus the palms require a mean annual temperature of 78°–81° F. and are associated with plantains and bananas, which are found in the moist places of equatorial regions, and form a staple item in the food supply of their inhabitants. The mimosa is noted to be absent in the temperate zone of the Old World, but found in the United States where 'in corresponding latitudes, vegetation is more varied and more vigorous than in Europe'. The heath form is widespread in the Old World. Arborescent heaths are found in the Atlas lands, Italy, and the south of Spain (apparently the dense scrub of the Mediterranean lands). The cactus, lianes, a 'tropical twining rope plant', and Gramineae are other typical forms (see 'Physiognomy of Plants', *Aspects of Nature*, vol. ii). In a second essay on steppes and deserts, the former are given a very wide interpretation, apparently any low open association. The heaths of the European lowland, from Jutland to the Scheldt, he characterizes as 'true steppe', and includes in the same category the prairies, llanos, and pampas of the Americas, and the grasslands of central Asia and Africa.

The doctrine of organic evolution, world-wide exploration, and research by natural scientists laid the modern foundations of biogeography. Robert Brown (1773–1858), through his acquaintance with Sir Joseph Banks, went out to Australia as naturalist to the *Investigator* in 1801. He brought back some four hundred plants, and later described the flora of the areas he had visited, and made comparisons with other regions in the southern hemisphere. Sir Joseph Hooker (1817–1911) went to the south seas with Ross, and from his researches there published his *Antarctic Flora*. In 1854 he published accounts of further biological researches in the Himalayas.

In the second half of the century efforts were made to relate plant distribution to environmental conditions. The chief stimulus was due again to German naturalists, Julius von Sachs, who published his *Handbuch der Experimental-Physiologie der Pflanzen* in 1865, and Haberlandt (*Physiologische Pflanzenanatomie*, 1884). A. Grisebach still assumed climate to be the principal factor in the distribution and character of plant life in his *Vegetation der Erde* (1875), while Engler in *Versuch einer Entwicklungsgeschichte der Pflanzenwelt* (1870–82) treated plant distribution from the evolutionary point of view.

In 1895 E. Warming of Copenhagen founded the study of plant communities in his *Plantesamfund* or *Ecology of Plants* (new edition, 1931). Shortly after appeared O. Drude's *Manuel de géographie botanique* (1897) (French trans.) and A. F. W. Schimper published his *Pflanzen-Geographie* (1898; Engl. trans., *Plant Geography*) on a new physiological basis. The ground was now prepared for further work on these lines, and much has been done in recent years. Quantitative methods of studying vegetation in relation to habitat were developed by F. E. Clements of the United States, and his method is expounded in *Research Methods in Ecology* (1905). This was the only work of its kind in the field until in 1923 Tansley published a *Practical Plant Ecology*, and in 1926 there appeared, under the joint editorship of Tansley and Chipp, *Aims and Methods in the Study of Vegetation*. The science of plant ecology is therefore a development of the twentieth century, and though much progress has been made, there is still a lack of precise knowledge of habitat factors, and the relation to them of plant life.

Plant ecology is literally 'the study of plants in their homes'. It comprises the study of the relations of plants, individuals, species, and communities, with their habitats. It has, therefore, two aspects, as defined by Warming: floristic plant geography, which is concerned with the distribution of plants or taxonomic groups, and ecological plant geography, the study of plant associations and their adaptation to habitat. The first is the study of species, the second of vegetation, and it is with the second that the geographer is primarily concerned. The 'habitat' consists of three elements, geographical position, determining the nature and variety of plants in relation to their past migrations; physical factors, which include the influence of soil (edaphic factors) and climate; and biological, i.e. the reactions of the plant on the soil.

According to their adaptation to the physical controls, Warming divided plants into four groups, and elaborated the idea and used the same terms suggested by Schouw early in the century.

> Hydrophytes: plants growing in soils with a watery substratum (over 80 per cent. water).
> Xerophytes: plants growing in dry soils.
> Halophytes: plants growing in soils with a high sodium chloride content in the substratum.
> Mesophytes: plants growing in soils which are not dry, wet, or salty.

Schimper recognized for the first time the important distinction between physical and physiological dryness or wetness. A soil may be humid, but the conditions are such that plants cannot absorb moisture—either through the acidity of the soil or low temperatures. On the basis of this physiological distinction he altered the above classification and adopted three main classes: Xerophytes, which inhabit physiologically dry soils; Hygrophytes, inhabiting physiologically wet soils; and Tropophytes, or plants which are hygrophytes and xerophytes at different seasons, e.g. deciduous trees and bulbous plants.

Schimper defined the essential aim of plant geography to be 'an inquiry into the causes of differences existing among

3490   Q

the various floras. The character of the vegetation on any part of the earth's surface is dependent on climatic factors—the main plant associations coincide with climatic controls—and local variations are due to edaphic factors.' But a third factor is of importance as determining the character of the flora, and the degree of its adaptation to its present habitat. 'Existing floras exhibit only one moment in the history of the earth's surface.' Plants are mobile and their present distribution is due to their dispersion, since their evolution in the Mesozoic era, from a circumpolar belt, which then had a sub-tropical climate, and in which there was free communication between the Old and New Worlds. The present character of world vegetation is a product of migration from this centre, the extermination of many species, and the local differentiation and multiplication of those which survived. In Europe east-west mountain barriers offered no escape for vegetation during the Ice Age, as opposed to North America, where the mountains are oriented from north to south, and where, as a result, plants could freely migrate and return with the withdrawal of the ice. Here lies the solution to Humboldt's unexplained observation—the greater richness of the flora of North America than of that of Europe in similar latitudes.

The distribution of plants may be treated floristically or ecologically. On the floristic basis we have the following world divisions (an elaboration of Drude's original classification):

SUB-REGIONS

NORTH TEMPERATE REGION:

1. Arctic Alpine.
2. Intermediate—the steppes of the Old World, the prairies of the New, and forests of both.
3. Mediterranean-Oriental. From the Mediterranean eastwards across the Old World, bounded to the north by the Caucasus and Hindu Kush.

4. China-Japanese.
5. Mexico-American bounded to the north by lat. 36°, and 40° on the Pacific Coast.

TROPICAL REGION:

1. African (with W. Arabia).
2. Indo-Malayan (Cochin-China, S. China, Malaya, Philippines, New Guinea, and Polynesia).
3. South American.

SOUTH TEMPERATE REGION:

1. South African.
2. Australasia (New Zealand, Tasmania, New Caledonia).

On the ecological basis Schimper's is the standard classification, on which all subsequent schemes are based. The basis of his scheme is climate, and the types of vegetation he regards as climatic formations. The world is divided into four belts, tropical, sub-tropical and warm temperate, temperate, and cold temperate or frigid. Each of these is then subdivided into vegetation regions as follows:

TROPICAL:

I. Woodland.	Rain Forest	Over 70 in. rainfall per annum.
	Monsoon.	
	Savanna.	
	Thorn.	
II. Grassland.	Savanna.	
	Steppe.	
III. Desert.	Scrub.	
	Succulent plants.	
	Perennial shrubs.	

WARM TEMPERATE:

I. Forests.	(1) Sclerophyllous—Evergreen Mediterranean.
	(2) Mediterranean Coniferous. (Maritime and Aleppo Pine.)
II. Steppes.	S. Algeria (Esparto Grass and Shotts).
III. Desert.	

COOL TEMPERATE:

   I. Forests. (1) Deciduous.
             (2) Coniferous.

COLD TEMPERATE or FRIGID:

   I. Dwarf Forests.
  II. Tundra.

As a part of the terrestrial environment vegetation is as vital to the geography as the form of the land, and several geographers have submitted works on general plant geography; while other works, based upon detailed research, are contributions of great significance to the complete study of the environment. Of the first type, the main contributions are Hardy's *Plant Geography* and D. H. Campbell's *Outline of Plant Geography* (1926). Of the latter, Tansley's *Types of British Vegetation* (1911) and, for a much larger area, Shantz and Marbut's *Vegetation and Soils of Africa* are representative works of their kind.

II

The geography of animals, though not developed, like plant geography, till the second half of the nineteenth century, had its beginnings in the second half of the eighteenth. Using the works of Buffon and Pallas, a German naturalist, Wilhelm Zimmerman, produced the first world map of the distribution of mammals (1777). In the early nineteenth century, as in plant geography, the distribution of vertebrates was assumed to be determined mainly by temperature. A. Wagner, following the same methods as Schouw in plant geography, delimited in the 'forties seven great faunal regions. Such maps also appeared in the Berghaus Atlas. It was early suggested that peculiarities of animal distribution were due to changes in the distribution of land and water. Thus Buffon speculated on a possible early connexion between Africa and South America, and Zimmerman suggested that the Sunda Isles were a former appendage of south-eastern Asia.

The main objects of zoogeography are threefold, first, to map the distribution of all forms of animal life; second, to delimit great faunal regions within which there are common and distinctive zoological features; third, to explain the phenomena of distribution. The pioneer in the delimitation of faunal zones in this country was P. L. Sclater, who, on the basis of the distribution of perching birds, submitted a scheme of six faunal regions to the Linnean Society in 1858. In 1860 Alfred Russel Wallace discussed the distribution of animals in Malaya and Australasia in an address before the same society, and in 1876 he enlarged his views in a book on the *Geographical Distribution of Animals*, which put the study on a sound basis. As a supplement to and completion of this work he published his *Island Life* in 1880. Wallace, whose researches were based on the distribution of mammals, accepted Sclater's regions.

Wallace's scheme of faunal regions is as follows:

1. PALAEARCTIC.   Europe to the Azores and Iceland; Asia north of the Himalayas and west of the Indus, with Japan and N. China; N. Africa (south to the Tropic of Cancer) and Arabia.

2. ETHIOPIAN.   Africa south of Cancer, and S. Arabia, Madagascar, and adjacent islands.

3. ORIENTAL.   (Indo-Malay).
India and Ceylon.
Indo-China.
South China.
Malay Archipelago, including Philippines, Borneo, and Java.

4. AUSTRALIAN—including the Pacific Islands, east to Marquesas.

5. NEOTROPICAL.   South America, West Indies, Central America, and Mexico.

6. NEARCTIC.   Temperate and Arctic North America, and Greenland.

Various modifications of this scheme, in nomenclature and in the addition of new regions and transitional regions, have been made by later zoologists. In 1874, from a study of the distribution of mammals, Sclater accepted a modification put forward by T. H. Huxley in 1868, that the name Arctogaea should be used to include the Nearctic, Palaearctic, Oriental, and Ethiopian regions. Huxley also suggested the term Notogaea to include the Australasian and Neotropical Regions. W. T. Blanford and H. Fairfield Osborn made further modifications, and Heilprin suggested the term Holarctic to include the Nearctic and Palaearctic, two transitional areas, one in Mexico and California (Sonoran Region) and the other the Mediterranean or Tyrrhenian (extending east across Western Asia) and a distinct Polynesian region. Later a Malagasy region was recognized. The last change adds the Celebes, Lombok, Flores, and Timor to the Australian Region.

The scheme of faunal regions in its final form is as follows (from Max Weber's *Die Säugetiere*, 1904):

I. ARCTOGAEA:

    1. Holarctic. 2. Ethiopian. 3. Malagasy. 4. Oriental.

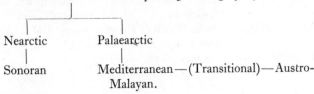

Nearctic   Palaearctic

Sonoran   Mediterranean—(Transitional)—Austro-Malayan.

II. NEOGAEA:

    5. Neotropical.

III. NOTOGAEA:

    6. Australian.
    7. Polynesian.
    8. Hawaian.

The distinctive features of the original six regions are summarized by Wallace as follows: 'These six regions, although all of primary importance from their extent, and well marked by their total assemblage of animal forms, vary greatly in their zoological richness, their degree of isolation, and their relationships to each other. The Australian region is the most peculiar and the most isolated, but it is comparatively small and poor in the higher animals. The Neotropical region comes next in peculiarity and isolation, but it is extensive and excessively rich in all forms of life. The Ethiopian and Oriental regions are also very rich, but they have much in common. The Palaearctic and Nearctic regions being wholly temperate are less rich, and they too have many resemblances to each other; but while the Nearctic region has many groups in common with the Neotropical, the Palaearctic is closely connected with the Oriental and Ethiopian Regions.'

The old idea of temperature as the main factor controlling animal distribution was naturally exploded by the new orientation to thought given by the doctrine of evolution, and by fossil remains found in all latitudes. Climatic factors are insignificant in the distribution and character of animals. Distribution is now recognized as being due to the varying rates and dates of migrations of animals, the most primitive forms having the widest distribution. Geographical barriers —mountains and water—are the chief factors directing the lines of migration, though their effectiveness is not equal for all forms of animal life. For mammals such barriers are absolute, and the presence of mammals in two separate land masses inevitably points to their former connexion. Owing to their inability to cross stretches of water, they are not found on islands—and in fact, as shown in Wallace's *Island Life*, islands, on account of their isolation, exhibit many peculiar features in the character of their fauna.

Chapter XIX

THE REGIONAL CONCEPT

TO the description of small areas Ptolemy gave the name 'chorography' as opposed to geography, which dealt with the earth as a whole, and 'cosmography' which dealt with the Universe. To him and to many of his successors geography and chorography were distinct, but to Nathaniel Carpenter the difference between the two was one of degree and not of kind, though he did not treat chorography in his work. Munster, on the other hand, paid no attention to 'geography' as defined above, but, like Strabo, described the countries of the world, their features, products, and the manners and customs of their peoples. Cluver adopted the opposite attitude. He defined geography as 'the description of the whole earth, so far as it is known to us' and he deals with the earth as a unit, not with its parts. Varenius, however, defined, in terms almost modern in conception, the scope and relations of geography or general geography, as he termed it, and chorography or special geography. Special geography should be approached through the application of general laws, and the description should fall under the heads of celestial, terrestrial, and human phenomena. 'These three kinds of occurrences are to be explained in special geography, and though the last sort seem not so properly to belong to this science, yet we are obliged to admit them for custom's sake and the information of the reader.'

At the beginning of the nineteenth century, as exemplified by the works of Pinkerton and Malte-Brun, special geography still consists of the unsystematic, encyclopaedic descriptions of countries. But a great advance was made by Carl Ritter, by treating in his *Erdkunde* not countries, but units or regions, into which he subdivided each of the continents. His method, however, was teleological and not based on scientific principles, and his descriptions lack systematization and correlation of human phenomena with natural conditions. Humboldt first applied the regional method of description with excellent results in his composite works on central Asia, Mexico, and the Llanos, and in his distribution of plants, but he did not suggest any regional divisions. No further progress had been made by the end of the nineteenth century, and it is interesting to note that British geographers were still playing with the term special geography, and one of the most prominent pioneers of British geography suggested the term chorography as the most suited to the detailed description of parts of the earth.[1]

Special, or, as it is now termed, regional geography, is thus a product of the last forty years. The culminating point, and as many would claim, the essential aim of modern geography, is thus the latest phase in the development of the subject. This anomalous feature must be attributed to the lack of criteria for the delimitation of areas, and of a method of description of human activities and their correlation with the physical environment. Progress in these two directions had to await first the collection, systematization, and cartographic representation of the distribution of natural phenomena on the earth's surface—winds, temperature, pressure, vegetation, cultivated products, &c.; and second, the refinement of the conception of the exact scope and aim of geography. Development in each of these directions has been traced in previous pages. Attention will now be turned to the development and nature of the modern regional method.

The regional concept received its first expression in France in the eighteenth century by Buache. Up to this time political divisions had always been accepted as the only basis for geographical description. As a result of Buache's theory of the arrangement of mountain systems, river basins were the first areas to be adopted in the delimitation of regions, on

[1] Cf. Pres. Address, Brit. Assoc., Section E, 1895.

the assumption that mountain ranges, or relatively high land, bordered every basin. More scientific treatment was attendant upon the preparation of geological and topographical surveys. Dufrenoy and De Beaumont produced a geological map of France, on the basis of which they proved the fallacy of Buache's theory. 'The geological lines which determine the form of the rocks define, as it were, the skeleton of a country, while the hydrographic lines only represent its purely external traits, which, on the same surface, change with time. Moreover, river valleys are only isolated furrows, whereas the general modelling of the relief of the earth is linked with geological features' (quoted by Gallois). This concept, however, found no echo, for it was far ahead of its time.

Modern regional geography had its founder in Vidal de la Blache, who in 1903 published his model *Tableau de la géographie de la France*, and in 1908 one of his pupils, Lucien Gallois, published a work on *Régions naturelles et noms de pays: Étude de la région parisienne*, in which he discusses the development of the regional idea. From other disciples followed a series of regional monographs, e.g. Demangeon, *Picardie* (1905); Blanchard, *Flandres* (1906); Vallaux, *La Basse Bretagne* (1907); J. Sion, *Les Paysans de la Normandie orientale* (1909); Levainville, *Le Morvan* (1909). With a fear of generalizations, and realizing the force of social organization and heritage in the human response to physical conditions, La Blache (*Annales de Géographie*, 1902) wrote: 'If this danger (of generalization) is to be feared, one must have recourse to antidotes. I can advise none better than the preparation of analytical studies, of monographs in which the relation between geographical conditions and social facts are viewed at close quarters, in a well-chosen and small field.' Again, Demangeon, in an appreciation of La Blache's regional method, writes: 'Every region has its unique character to which contribute the features of the soil, atmo-

sphere, plants, and man. The aim of all research consists of
the analysis of these features. The aim of description is
to synthesize them, and to show the interlocking of all the
phenomena which comprise regional types' (*Revue Universitaire*, 1918). Brunhes also advocated and practised the
regional method. He considered that 'it is a wise principle
of method and introduction to limit oneself, at the beginning, to determining the geographic connexions between
natural facts and human destinies through the medium
of the "essential facts" analysed in small natural districts'.
A series of such studies, illustrating his method, are contained
in his *Human Geography*.

For Gallois, in his regional division of the Paris Basin,
a region must have 'une impression d'ensemble'. Climate,
altitude, and structure should be considered in regional delimitation, and a region so defined will be a 'natural' region,
since it is based on natural features. But climatic variations
are only marked over large areas of the earth's surface;
structural variations occur frequently over small areas and
give rise to distinct local contrasts of land form and vegetation. Hence, in defining small units, subdivisions of the Paris
Basin for example, structure should be the main criterion.

Regional study was developed to a small extent in Germany
during the last years of the nineteenth century. A 'Central
Commission on the Regional Geography of Germany',
established in 1886, encouraged the preparation of monographs which appeared under the title of *Forschungen zur
deutschen Landes- und Völkerkunde*, edited by R. Lehmann,
and after 1888 by A. Kirchhoff. Kirchhoff also edited the
Länderkunde von Europa (1887–93) to which contributions
were made by Penck, Supan, Fischer, and Lehmann. But
these studies lacked the synthetic method inculcated by La
Blache of treating human adjustment to environment. They
were more specialized, as for example 'The Towns of the
North German Plain, in relation to the configuration of the

ground' by Dr. Hahn, 'The Plain of the Upper Rhine and its Neighbouring Mountains' by R. Lepsius, 'Mountain Structure and Surface Configuration of Saxon Switzerland' by A. Hettner, 'The Erzegebirge: an Orometric-Anthropo-geographical Study' by J. Burgkhardt. Penck in his *Morphologie der Erdoberfläche* first drew attention to the smallest units of land which have similarity of topographic form. To such small regions he gave the name 'Landschaft'; and thus he wrote, for example, of Moränenlandschaft—i.e. regions or landscapes of morainic areas,—dune regions, and volcanic regions.

At the close of the century a scheme had been suggested and begun for a small area, in the south-west of Sussex, by H. R. Mill, for the systematic description of the one-inch Ordnance Survey maps of England and Wales, on the same lines as the geological survey reports[1]. The scheme did not materialize, but was adopted by Mackinder, at the University of Oxford, and later by the heads of other departments of geography, by allotting sheets to students for dissertation work: three of these early studies were published. The first published study of the small regional divisions of a part of England, comparable to and, in fact, inspired by the French school's method, was Roxby's treatment of the 'Historical Geography of East Anglia' which appeared as two articles in the *Geographical Teacher* (1907–8). He defines a natural region as 'an area throughout which a particular set of physical conditions prevail, and *ceteris paribus*, a particular set of physical conditions will lead to a particular type of economic life. A physical unit tends to become an economic unit.' On this basis he gave the first accurate delimitation of minor regions within one of the larger regions of the British Isles.

It will be obvious from the foregoing, that attention was first directed to the smallest geographical units, defined on

[1] *Geog. Journ.* vol. xv.

the basis of their structure and topography. But there exist larger regions, of various orders of magnitude, in the delimitation of which different factors of the physical environment must be adopted as criteria. In the first decade of the present century two schemes were put forward for the division of the world into its component natural regions, on the basis of physical criteria, the first by A. J. Herbertson of Oxford, and the second by a German, and a lifelong devotee of geographic method, A. Hettner.

Physical environments differ throughout the world owing mainly to differences of relief, climate, vegetation, and in order to interpret the human response to them, these natural phenomena should be synthesized, and regions demarcated, each with a similar type of natural environment. Such a scheme for the world, based on general climatic distributions, was prepared by Herbertson in 1904, and published in the *Geographical Journal* (vol. xxv, 1905). The task of regional delimitation involved two main problems, to one of which Herbertson offered a solution.

(1) What criteria should be adopted in their delimitation?
(2) How are the different orders of natural regions to be determined?

'A natural region', he argued, 'should have a certain unity of configuration, climate, and vegetation. . . . The mapping of human conditions has less significance in indicating the natural geographical regions, for the factor of human development has to be taken into account as well as the possibilities of the natural environment. Political divisions expressing the more complex and comparatively unstable human conditions must be eliminated from any consideration of natural regions.' At the Oxford School of Geography, Herbertson, with the assistance of students, prepared world maps to show structural regions, thermal regions, and the distribution of rainfall and vegetation, now published in the Oxford Wall

Map Series. From these distributions he prepared his classification of Major Natural Regions, which, as Roxby justly maintains, 'whatever criticisms may be directed upon it, represents one of the most fruitful and constructive achievements in the development of modern geography'.

The world is divided into temperate belts, on the basis of the thermal regions—Polar, with no month with a temperature over 50° F.; Cool Temperate, roughly between latitude 40° N. and S. and the polar circles; Warm Temperate, approximately between latitudes 30° and 40°; and the Hot Belt with temperatures over 68° F. most or all of the year. The subdivisions of each of these belts is based mainly on rainfall distribution and partly on relief. The classification is as follows:

1. POLAR REGIONS:
   (a) Lowlands (Tundra).
   (b) Highlands or Ice Caps (Greenland).

2. COOL TEMPERATE REGIONS:
   (a) Western Margin of West European Type.
   (b) Eastern Margin or St. Lawrence Type.
   (c) Interior Lowlands or Siberian Type.
   (d) Interior Highlands or Altai Type.

3. WARM TEMPERATE REGIONS:
   (a) Western Margin or Mediterranean Type.
   (b) Eastern Margin with Summer Rains or China Type.
   (c) Interior Lowlands or Turan Type.
   (d) Plateau or Iran Type.

4. HOT REGIONS—TROPICAL:
   (a) Western Desert or Sahara Type.
   (b) Monsoon Summer Rain Type.
   (c) Summer Rain Type of interior or Sudan Type.

5. LOFTY TROPICAL or SUB-TROPICAL MOUNTAINS:
   Tibetan Type.

6. EQUATORIAL:
   Wet Equatorial Lowland or Amazon Type.

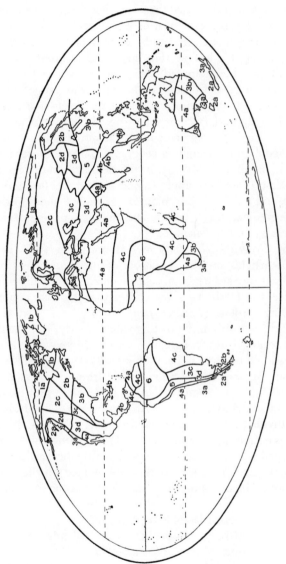

Fig. 30. Herbertson's Natural Regions.

Herbertson slightly varied these divisions in other versions. He also realized the existence of minor regions of varying orders, 'organs, tissues, and cells' within the 'macro-organism of the earth', but at that stage, he asserted, definitions could not be attempted, although he instanced W. M. Davis's 'Drainage of Cuestas' as the type of investigation he had in mind. It is of interest to read the discussion at the Royal Geographical Society following Herbertson's paper. It met with a lukewarm reception. 'Chorography', it was maintained by an outstanding geographer, 'would never take root', and another argued that in the delimitation of regions of the first order, orography alone should be taken into account, and a third of our pioneers said, 'we should stick to the degree net, the coast line and the contour line, which are necessary and sufficient.'

In the *Geographische Zeitschrift* (Leipzig) in 1908, in a series of articles on the 'Division of the Lands', Alfred Hettner put forward a second scheme on an entirely different basis. He, like Herbertson, recognized that there are regions of varying orders (*Landschaften*, *Lande*, and *Erdteile*), which have not only to be distinguished clearly, but should also be arranged in a progressive series. Moreover, all the regions should be natural, i.e. they should take no cognizance of human factors. But Hettner differs from Herbertson in that, working down from the largest regions, he first makes the natural distinction between land and water. The former is then divided into continents and islands. The islands are classed as oceanic and continental, and the continental islands are grouped with the respective regions of the first order, the continents. Each continent is then subdivided on the basis of position, climate, and structure. Thus Europe is divided into—

(1) The northern islands and peninsulas.
(2) The south European peninsulas and islands.
(3) The continental block (*Kontinentalrumpf*).

Of the subdivisions, Italy, for example, is divided into Upper Italy, the peninsula, and the islands; and the British Isles are divided on a political basis. The third large region is divided into three units, France, central Europe, and the East European Plain (see *Grundzüge der Länderkunde*, Bd. I, Europa, 1907).

Asia is divided into five major natural regions as follows:

Northern Asia (Siberia).
  Western Siberia.
  Eastern Siberia.

The Near East.
  Aral-Caspian Depression.
  Folded Mountains and from Asia Minor to Iran.
  Syrian-Arabian Massif.

Central Asia.
  Tarim Basin.
  Mongolia.
  Tibet.

Eastern Asia.
  Okhotsk with Kamchatka and Kuriles.
  Amurland, Manchuria, and Japan.
  China with Formosa and Liu Kiu Isles.

Southern Asia.
  India.
  Further India.
  East Indies.

These two continental subdivisions serve to illustrate the method of the scheme, and the way in which it differs from Herbertson's. The essential difference is that the one takes the continents and the other the world as a basis: the world view, and the grouping of regions with similar environments, is not obtained by Hettner's continental method.

During the last twenty-five years research on regional

method has been grappling with two main problems. First, how far should the modification of the physical environment be considered in the delimitation of regions? Secondly, on what criteria must the delimitation of regions of varying orders be based, and how shall they be grouped together? The two are intimately connected, and their solution is dependent on intensive regional studies.

In England the theory of natural regions has been expounded by J. F. Unstead and P. M. Roxby. Unstead produced the second scheme of natural regions in 1910, its basis of world classification of natural environments being similar to Herbertson's. Six years later he suggested (*Geographical Journal*, 1916) a synthetic method of delimiting regions, which would take into account both physical and human elements.

He sums up as follows:

(1) The present delimitation of natural regions is based upon physical conditions; geographical regions should be delimited, taking into account the human equally with the physical factors.

(2) The present method depends upon certain analysed elements such as isotherms and isohyets of various kinds; the geographical units should be obtained as far as possible by considering the synthetic effect of these and other elements.

(3) The geographical units should be determined as areas having common predominant characteristics; these and the less noticeable features should be recorded accurately and, as far as possible, quantitatively.

(4) The method hitherto adopted analyses the world into large divisions; the proposed method is to proceed synthetically by building up the larger regions from smaller areas already delimited and characterized.

This synthetic scheme may be regarded as the goal of

geography, but it is attendant upon detailed work on the smallest units, and definite criteria for the larger.[1]

Roxby in his article on 'The Theory of Natural Regions' (*Geographical Teacher*, 1926) has also applied his method to the regional subdivision of Europe, and in particular, central Europe. He writes: 'It is the comprehensive study of the region and of inter-regional relations which gives unity and distinctiveness to geographical investigation, and the region as so conceived may be compared to an organism, at least as implying a complex entity made up of a particular integration of different elements, physical, biological, and human.' The major natural regions ('geographical' according to Unstead) should combine 'a distinctive association of intrinsic conditions with a definite set of space relations', and the lower categories of regions should be marked by the prevalence of a particular character or relations, the smaller the units the narrower the basis of differentiation.

The regional method in geography has been the subject of serious thought and investigation on similar lines by a number of German geographers, particularly Hettner, Passarge, Grano, and Braun. They generally recognize regions of several orders, all of which are essentially natural regions, based on criteria of land form and vegetation for the minor regions, with the addition of climatic conditions and space relations for the larger regions. The smallest distinctive geographical unit, named *chore* by Solch and *Landschaftsteil* by Passarge, consists of the 'elements of surface distribution' and may be only a few square miles in extent. A number of these units grouped together constitute a *Landschaft*, a region characterized by distinctive features of topography and vegetation. The next region, the *Land-* or *Landesteil*, consists, according

[1] The working of the method has been shown by Unstead in his essay on the regions of Spain (*Scottish Geog. Mag.* 1917) and their grouping into regions of different orders, culminating in the North-West European and Mediterranean major geographical regions, with a group of transitional regions between the two in central Spain.

to Penck and Hettner, of a number of separate *Landschaften* 'which on account of their juxtaposition . . . have mutual relations and a physiognomic unity, and taken together form a geographical unit' (Penck, *Deutschland als geographische Gestalt*). Examples of these larger units of Penck's scheme are Great Britain and Ireland, and Fennoscandia in the north of Europe, the three peninsulas of the Mediterranean, and France, Germany, and the Danube Basin. The basis of this scheme is the Continent, and the larger subdivisions thereof are similar to those suggested by Hettner. Passarge, on the other hand, groups the *Landschaft* of which the Baltic Heights may be taken as an example, to form *Landgebiete* (e.g. the North German Lowland) which possess certain features of physical unity. These again form the component parts of larger climatic and vegetational regions (e.g. Central European region) which are members of a general comparative world scheme of regions, classified on the basis of natural conditions. Both Braun and Passarge consider the goal of regional geography to be the explanation of the transformation of a natural to a cultural region or 'landscape'.

It will be apparent that regions, areas of the earth's surface possessing a certain unity of physical or human conditions—or both—may be drawn up to show the distribution of single physical characteristics, such as structure, climate, or vegetation, or these may be combined to form natural regions. There is a general disinclination to include human factors in regional delimitation; the accepted method is to take physical criteria. It is impossible to combine all elements in a world classification, since man and nature do not obey the same laws—'understanding is more than classification'. The prime aim of the geographer is to interpret the character of human adjustment to the natural environment. The variations of environment, the outcome of physical factors, distinguish the natural regions as a framework for the investigation.

Regions of all kinds have been suggested to show the

distribution of various phenomena. The United States has been divided into physiographic regions by American geographers. On the other hand, regions of human activities have also been prepared and studied. Fleure has made a valuable contribution to human geography in his concept of human regions, based upon the main features of human life in response to similar natural conditions (*Scottish Geog. Mag.*, 1919). Thus there are regions of debilitation as in the equatorial forests, regions of effort as in the lowlands of north-western Europe, regions of difficulty in highland areas, regions of increment as in the fertile parts of the Mediterranean, regions of privation as for example in the tundra, and regions of wandering as in the grasslands of central Asia. In the sphere of economic geography agricultural regions of different orders are being thoroughly worked out for each continent in the American periodical *Economic Geography*. C. B. Fawcett has marked out for England and Wales regions within which one great city dominates economic, social, and administrative activities. Such areas, it is claimed, should serve as a more rational basis for the reorganization of administrative areas. Regions of a similar kind have been worked out in France, the home of the regional movement, where, on the geographical side, it was sponsored by La Blache. Some years ago De Preville submitted a map of regions for Africa, based upon community of social organization and activities. In more recent years Clark Wissler has demarcated the cultural regions of pre-Columbian America.

While all these schemes of regions, showing the distribution of various phenomena, are of importance, the essence of geography is the explanatory description of human occupance within composite natural regions. While the principles of method are clearly defined, little work has yet been done on its detailed application. Herein lies a main future task of the geographer.

Chapter XX

SUMMARY AND CONCLUSION

GEOGRAPHY, till the nineteenth century, was considered to be a description of the earth—the description of all terrestrial phenomena. Hence, it had three distinct aspects. First, the earth as a member of the universe; second, the earth as a unit—its shape, size, and arrangement (and methods of determining) of latitude and longitude; and third, the detailed description of its component parts on the basis of political divisions. To these three aspects Ptolemy gave the names of cosmography, geography, and chorography, though in his view they were not branches of one science, but entirely separate subjects, dealing with different groups of related facts. Ptolemy, as we have seen, dealt with geography mainly on its mathematical and astronomical sides, whereas Strabo described countries in great detail and established the method which was followed centuries later with the Renaissance revival. The Greek scientists, led by Ptolemy, established mathematical geography on a firm foundation, in spite of the lack of knowledge of the earth, the lack of data, and the difficulty of measurement.

After a long period of the Dark Ages in Europe, when the Greek views were occasionally echoed in a wilderness of religious mysticism, the Renaissance witnessed a great revival of thought, with the revival of classical Greek works, and the expansion of knowledge of the globe. The sixteenth and seventeenth centuries are marked by world-wide exploration, the discovery of new lands and facts, and the first positive proof of the rotundity of the earth. The medieval cosmographers were faced with the problem of mapping the new lands of the spherical earth on a new plane surface. Ptolemy's records of latitude and longitude were collected, corrected, and, as far as the inaccurate methods of the time

would permit, amplified; while new projections early displaced Ptolemy's, which was suited only to the mapping of
one hemisphere.

Advances were made in local, as opposed to world, cartography, for the method of triangulation, with compass and
plane-table, came into use in the sixteenth century, for the
preparation of detailed surveys.

Yet the distortion of the distribution of land and water,
the reliance on Ptolemy's Far Eastern records, in the absence
of others more reliable, retarded accurate cartography. Mercator emancipated himself to some extent from Ptolemaic
tradition, but this was not finally achieved till the eighteenth
century, when the chronometer and sextant were in general
use, methods of accurate land survey were invented, and
many more records had been collected, particularly by the
Jesuits in the Far East. Delisle and D'Anville laid the foundations of modern cartography.

During the sixteenth, seventeenth, and eighteenth centuries, facts and phenomena were noted and recorded by
explorers and traders, and also, mainly on the mathematical
side, at the instigation of the French Academy of Sciences
and the British Royal Society of Arts. Scientific exploration,
however, begins at the end of the eighteenth century with
Cook's voyages to the South Seas, which finally laid low the
centuries-old idea of a vast southern continent—though, on
the other hand, he circumnavigated a new island continent,
Australia.

Cook's voyages herald the period of extensive and pains-
taking scientific exploration, while this was accompanied by
empirical attempts to co-ordinate the new data. At the same
time the early nineteenth century witnesses the dawn of
modern scientific thought, and the birth of the concept of
evolution, mainly through the early works of Lamarck and
Laplace, and later, on its inorganic and organic sides, by
Lyell and Darwin respectively.

In the realm of geography Germany is to the fore throughout the century, with its giants in Humboldt and Ritter. They first introduced the principles of causality and co-ordination into geographical method, through the comparative study of similar phenomena in various parts of the earth's surface. The latter developed further the principle of co-ordination, and established the concept of regional treatment. Ritter first offered a method of systematic treatment of distinct regions of the earth, viewed as parts of a whole (i.e. the continents), based on the interrelations of man and nature.

The empiricism of these two pioneers is replaced by rational methods in the second half of the century, on the physical side by Peschel and his successor Richthofen, and on the human side, by Ratzel and Leplay and their disciples. To Ratzel is principally due the third principle of geographical method—the principle of extension and distribution.

The last third of the nineteenth century witnessed great strides in geography, particularly in Germany, a fact which is to be associated with the rapid extension of inland exploration—especially in Africa—and is reflected in the foundation of new scientific periodicals. The older geographical societies were founded mainly for the advancement of exploration—the geographical societies of Paris (1825), Berlin (1827), and London (1830). A later product of exploration in the latter half of the century is the periodical *Petermanns Mitteilungen*. But new scientific periodicals, concerned with the advancement of academic geography, were established at the end of the nineteenth and the early years of the twentieth century, e.g. *Geographische Zeitschrift*, *Annales de Géographie*, *Bulletin of the American Geographical Society*, *The Geographical Teacher*. Meanwhile, modern geography rapidly grows to its modern form. The genetic interpretation of land forms is studied by a body of geologists, of whom Richthofen, Penck, and Davis are the main contributors; while the

elaboration and refinement of the Ratzelian anthropogeography is taken up by Vidal de la Blache in France.
Britain still lagged behind, though it had pioneers in
Mackinder, Chisholm, Keltie, and Mill, who drew upon
German sources for their chief inspiration. And one great
contribution came from the ranks of British geographers—
the concept and elaboration of a world scheme of natural
environments or regions, by Herbertson.

The regional concept is a product of the twentieth century
and its development coincident with what Hettner has called
the modern phase of 'shrinkage and synthesis'. Owing to the
nature of geography the number of its definitions are legion,
but the idea prevalent at the end of the nineteenth century
was that it consisted of two distinct aspects, physical and
human. The difficulty of defining its relations to the kindred
sciences led to protest from some of these, mainly geology
and sociology. The dualistic concept was partly defeated
by Ratzel—though in his wide interpretation of human
geography, he overlapped into other sciences, and broached
problems which are still unsolved by specialists in them—
while Richthofen made, in 1883, a clear definition of the interdependence of the physical, human, and biological aspects
of geography.

One of the definitions of geography, particularly current
some twenty years ago, regarded it as the science of distribution.[1] Man was regarded as the culminating point of the
subject, but it was equally concerned with all phenomena
on the earth's surface. The subject still wears a coat far too

[1] Herbertson defined the subject as concerned with the distribution, not
of one element, but of all. Mill similarly claimed it to be the the science
which deals with the forms of relief of the earth's crust and with the influence which these forms exercise on distribution of all other phenomena—an attitude which is reflected in the make-up of the *International
Geographies* (1899) edited by Mill. W. M. Davis has defined geography
as 'modern geology' and suggested the term ontography for the branches
which deal with the distribution of organic life in its relations with the
earth.

big for it. It is now usually defined as the study of the 'inter-relations between man and his environment'—a very vague concept. Where does it end? In interpreting environmental relationships facts must be drawn from kindred physical and social sciences. Thus, we have climatology (based on meteorology); physical geography, based on geology; mathematical geography, based on astronomy on the physical side; while on the human side, racial geography is dependent on anthropology and psychology, economic geography on economics, social geography on sociology, political and historical geography on history, and finally biogeography on zoology and botany. As an illustration of the enormous field of the subject, historical geography alone deals with the relations of man, in his evolution, with environment. In other words, geography, taking the whole of history as its field, selects therefrom those facts or trends which illustrate the dependency of man and his reaction on physical conditions. Again, the geography of animals deals with the origin, migrations, and present distribution of animal life in all its forms, from mammalia downwards. Racial geography is concerned with problems which are far removed from the relations of man and his environment, or is concerned with relations of this nature which are at best merely tentative. It is obvious that the field is still too great for one science. Much of its peripheral field, which deals either with problems which could definitely be better tackled by a specialist in the other camp, or are geographical merely in that they deal with distribution, should be evacuated. There is still ample scope for further shrinkage, a fact which is evidenced by the still frequent articles on its aims.

In the endeavour to limit the field there now exist two schools of thought, the one concerning itself with human adjustment to the physical environment—the view generally held in Britain—and the other with the description and interpretation of the physical and cultural landscapes.

The systematization, reasoned synthetic description, and correlation of the physical and human facts gave birth to the regional concept—the chorography of Ptolemy and later writers. It was to the elaboration of this concept that German geographers, led by the veteran pioneer Hettner, Vidal de la Blache, and A. J. Herbertson, directed their attention. It is, or should be, the final goal of modern geography, its core and culmination.

The distinctive sphere of geography is the region within which framework is played the drama between man and environment. The study of the region is its distinctive field, untouched by any other social or physical science, and herein lies the goal of modern geographical research. Human geography is concerned with man's relations to environment and in the biological field, facts of organic life and distribution are of no consequence to the geographer except in so far as they constitute part of the physical environment with which man has relations. Geography must draw upon kindred sciences, as Ritter pointed out, and weave the derived facts into the network of its regional framework.

Thus, while La Blache established the method of regional description on a local basis, and Herbertson the concept of major natural regions on a world basis, the main credit for interrelating the two and analysing exhaustively the method of treatment, lies with the modern school of German geographers, who thus carry on the tradition of their predecessors.

It will not be inappropriate to make a few concluding remarks on the development of geography in education.

In 1886 John Scott Keltie submitted a report to the Royal Geographical Society on geography in education in Britain and on the Continent, with a view to its promotion in this country. In most countries of the Continent geography at that time was taught in schools of all grades, and on the same level as other subjects and, moreover, it was taught by teachers trained in geography. Finally, geography was recognized in nearly all the universities. There were twelve professors in German universities mainly established since 1870 —before that date there had only been one at Berlin, originally held by Ritter. In Germany also the 'facultas docendi' in geography was a degree awarded to intending teachers in gymnasia and 'Realschulen'. In 1871[1] no geography was taught in the schools of France. In 1886 it was a recognized subject, and there were chairs at most of the universities, and normal schools for teachers. In Italy, geography was taught at twelve universities, its development being due to Giuseppe Dalla Vedova (1834–1919) and Giovanni Marinelli (1846–1900).

In England, though some improvement had been effected in elementary schools, geography elsewhere was non-existent.

'In our secondary schools, including the great public schools, with rare exceptions, depending mainly on the attitude of the master in charge of the subject—if there was one—geography was treated as a poor relation, "neglected and despised", mainly consisting of lists of names to be committed to memory. In the universities it was practically unrecognized, either on the side of science or on the human side. So it was in the civil service examinations and in the training institutions for the army and navy. The text-books, maps, and appliances available for teaching purposes were on a level with the position of the subject in education' (Keltie).

The Royal Geographical Society had been endeavouring since 1871 to secure the recognition of geography at Oxford and Cambridge. This was not achieved, however, till after the preparation of Keltie's report, at Oxford in 1887 and at Cambridge in 1888. Now geography is taught at most university institutions, and of these most have honours degrees in arts and science.

[1] Levasseur, *L'Étude et l'enseignement de la géographie* (Paris, 1872).

Taking a broad view, the development of geography down to modern times is a product of the Continent and particularly of Germany, with the rich tradition of Humboldt and Ritter. It is now a definite body of knowledge, with a definite aim and a definite method. It has extensive and vague peripheral fields, but its goal is the region, the most recent product of its development. It is the region which crystallizes the synthesis of human and physical elements which is its central theme. Not only does geography fill a place in the realm of the social sciences, but it has inestimable value as a discipline and as a source of sympathetic understanding of world problems. It is this last aspect which gives it great value in education in the modern world.

BIBLIOGRAPHY

The bibliography is a selected list of books and articles which deal specifically with the history of geography. Standard works, new and old, are mentioned in the text.

GENERAL

Hettner, A. *Die Geographie: Ihre Geschichte, ihr Wesen, und ihre Methode* (Breslau, 1927).

Wisotzky, E. *Zeitströmungen in der Geographie* (1897). (Mainly eighteenth and nineteenth centuries.)

Peschel, O. *Geschichte der Erdkunde bis auf Humboldt und Ritter* (1865).

Saint-Martin, Vivian de. *Histoire de la géographie* (Paris, 1873). (Mainly cartography.)

Wagner, H. *Lehrbuch der Geographie* (Leipzig, 1900). (Introduction, vol. i.)

Baker J. N. L. *A History of Geographical Discovery and Exploration* (London, 1931).

For general development since 1877 see *Geographisches Jahrbuch.*

CHAPTERS I–X. EARLY HISTORY

(In addition to works cited above.)

Bunbury, Sir E. H. *History of Ancient Geography* (London, 1879).

Beazley, C. R. *The Dawn of Modern Geography* (Oxford, 1897–1906).

Tozer, H. F. *A History of Ancient Geography* (Cambridge, 1897).

Berger, H. *Geschichte der wissenschaftlichen Erdkunde der Griechen* (Leipzig, 1891).

CHAPTER XI. GERMAN SCHOOL

Gallois, L. *Les Géographes allemands de la Renaissance* (1890).

Gunther, S. 'Peter und Philipp Apian: zwei deutsche Mathematiker und Kartographen: Ein Beitrag zur Gelehrten-Geschichte des XVI. Jahrhunderts', *Abhand. der Königl. Böhm. Gesellschaft der Wissenschaften*, VI. Folge, II. Band (1882).

Kretschmer, K. 'Die physische Erdkunde im christlichen Mittelalter', *Geog. Abhand.* (Penck), Band 4, Heft 1 (1889).

Park, G. B. *Richard Hakluyt and the English Voyages*, Research Series, Am. Geog. Soc. (1928).

Taylor, E. G. R. *Tudor Geography* (1931), also articles in the *Geographical Journal*.

Encyclopaedia Britannica. Articles on 'Geography' and 'Map'.

Baker, J. N. L. 'Nathaniel Carpenter and English Geography in the Seventeenth Century', *Geog. Jour.*, vol. lxxi (1928).

Cross, W. R. 'Dutch Cartographers of the Seventeenth Century', *American Geog. Rev.*, vol. v (1918).

Wauermann. *L'École cartographique belge et anversoise du XVIe siècle*, 2 vols. (1895).

CHAPTER XII. FLEMISH SCHOOL

Raemonk, J. van. *Gérard Mercator, sa vie et ses œuvres* (1869).

Partsch, L. G. 'Philipp Cluver; der Begründer der historischen Länderkunde; Ein Beitrag zur Geschichte der geographischen Wissenschaft', *Geog. Abhand.* (Penck), Band 2, Heft 5 (1891).

Gallois, L. 'La Géographie générale de Varenius', *Journal des Savants*, Nouv. Sér. iv (1906).

CHAPTER XIII. MEASUREMENT, CARTOGRAPHY
(1650–1800)

Heawood, E. *Geographical Discovery in the Seventeenth and Eighteenth Centuries.*

Gunther. *Early Science in Oxford* (8 volumes).

Curnow, I. J. *The World Mapped* (1930).

Reeves, A. E. *Maps and Map Reading* (1910).

Reeves, A. E. 'Mapping of the Earth. Past, Present and Future' *Geog. Jour.*, vol. xlviii (1916).

Taylor, E. G. R. Articles in *Geog. Journal.*

CHAPTER XV. HUMBOLDT AND RITTER

Gage, W. L. *Life of Carl Ritter* (1867).

Bruhns, K. *Alexander von Humboldt*, 3 vols. (1872). Translated by Lassell (1873).

Marthe, F. 'Was bedeutet Karl Ritter für die Geographie?', *Zeit. der Ges. für Erd.* (1879).

Hozel, E. 'Das geographische Individuum bei Karl Ritter und seine Bedeutung für den Begriff des Naturgebietes und der Naturgrenzen', *Geog. Zeit.* 2 (1896).

Vidal de la Blache. 'La Principe de la géographie générale', *Annales de Géographie*, iv (1896).
Hettner, A. 'Die Entwicklung der Geographie im 19. Jahrhundert', *Geog. Zeit.* 4 (1898). (See also (9) 1903 and (11) 1905.)
Mackinder, H., and Mill, H. R. Presidential Addresses to Section E, British Association, 1895 and 1901.

CHAPTER XVI. PHYSICAL GEOGRAPHY

Mehedenti, S. 'La Géographie comparée d'après Ritter et Peschel', *A. de Géog.*, x (1901).
Davis, W. M. 'Geographical Essays (History of Development of Geomorphology). Progress of Geography in the United States', *Annals of Assoc. Am. Geog.*, Dec. 1924.
Zittel, Karl von. *History of Geology and Paleontology* (1901).
Shaw, Napier. *Manual of Meteorology*, vol. i. Meteorology in History.

CHAPTER XVII. HUMAN GEOGRAPHY

Febvre, L. *Geographical Introduction to History* (1925). For the contrasted methods of Determinism and Possibilism see Semple's *Influences of Geographic Environment*, and de la Blache's *Principles of Human Geography*.
Dryer, C. R. 'Genetic Geography', *Annals Assoc. Am. Geog.*, vol. x.
Schrader, F. *Foundations of Geography in the Twentieth Century*. First Herbertson Memorial Lecture.
Brunhes, J. 'Human Geography', in *History and Prospects of the Social Sciences*, ed. by H. E. Barnes (New York, 1925).
——. 'Friedrich Ratzel', *La Géog.*, x (1904).
Richthofen, Ferd. von. *Aufgaben und Methoden der heutigen Geographie* (1883).
Vidal de la Blache. 'Les Conditions géographiques des faits sociaux', *A. de Géog.* xi (1902); 'Les Caractères distinctifs de la géographie', *A. de Géog.* xxii (1913); 'La Géographie politique d'après Ratzel', *A. de Géog.* vii (1898).
Raveneau, L. 'L'Élément humain dans la géographie' (Summary of Ratzel's *Anthropogeographie*), *A. de Géog.* i (1891).
Huckel. 'La Géographie de la circulation selon Ratzel', *A. de Géog.* xv (1906); xvi (1907).
Durkheim. 'Review of Ratzel's *Anthropogeographie*', *Année Sociologique*, iii (1898–9).

Demangeon, A. 'Vidal de la Blache', *Revue Universitaire*, June 1918.

Haddon, A. C. *History of Anthropology*.

Eckert, M. 'Friedrich Ratzel als Akademischer', *Geog. Zeit.*, vol. xxv (1919).

Dietrich, B. 'Alexander Supan', *Geog. Zeit.*, vol. xxiv (1918).

Philippson, A. 'F. von Richthofen als Akademischer', *Geog. Zeit.*, vol. xxvi (1920).

Ruhl, A. 'Theobald Fischer als Methodiker der Geographie', *Geog. Zeit.*, vol. xxvii (1921).

Steffen, H. 'Alfred Kirchhoff als Methodiker der Geographie', *G.Z.*, vol. xxv (1919).

Gallois, L. 'P. Vidal de la Blache', *A. de Géog.* xxix (1918).

Lamprecht, K. *Friedrich Ratzel*.

Sorokin, P. *Contemporary Sociological Theories* (1929).

Sauer, C. O. 'Cultural Geography' in *Recent Developments in the Social Sciences*, ed. E. C. Hayes, Lippincott's Sociological Series, Philadelphia (1928).

CHAPTER XVIII. BIOGEOGRAPHY

Refer to standard works given in text and *Encyclopaedia Britannica*, articles on 'Zoology' and 'Plants' (13th edition).

Beddard, F. E. *A Text Book of Zoogeography* (1895).

CHAPTER XIX. REGIONAL CONCEPT

Herbertson, A. J. 'The Major Natural Regions', *Geog. Jour.*, vol. xxv (1905); 'The Higher Units', *Scientia* (1913).

Unstead, J. F. 'A Synthetic Method of Determining Geographical Regions', *Geog. Jour.* (1916).

Roxby, P. M. 'The Theory of Natural Regions', *Geog. Teacher* (1925).

Passarge. *Die Grundlagen der Landschaftskunde* (3 vols.); 'Die natürlichen Landschaften Afrikas', *Pet. Mitt.* liv (1908).

Hettner, A. *Die Oberflächenformen des Festlands: Ihre Untersuchung und Darstellung* (1921).

Vidal de la Blache. 'Des Divisions fondamentales du sol français' in *La France*, vol. i of *Cours de Géographie* (1897).

Gallois, L. *Régions naturelles et noms de pays* (1907).

Penck, A. 'Neuere Geographie', *Zeit. der Gesell. für Erd.* (1928).

Sauer, C. O. 'The Survey Method in Geography and its Objectives', *Annals Assoc. Am. Geog.* (1924).

Gradmann, R. 'Das harmonische Landschaftbild', *Zeit. der Gesell. für Erd.*' (1924).

Whitbeck, R. H. 'Geography in American and European Universities', *Journal of Geography*, vol. xviii.

Keltie, J. Scott. *Position of Geography in British Universities* (1921) (Am. Geog. Soc., Research Series).

De Martonne, E. *Position of Geography in France* (Am. Geog. Soc., Research Series).

Joerg, W. L. G. 'Recent Geographical Work in Europe', *Am. Geog. Rev.*, vol. xii.

Davis, W. M. 'An Inductive Study of the Content of Geography', *Annals Assoc. Am. Geog.* (1905).

INDEX